To Have and Have Not

To Have and Have Not

Energy in World History

Brian C. Black

ROWMAN & LITTLEFIELD
Lanham • Boulder • New York • London

Published by Rowman & Littlefield
An imprint of The Rowman & Littlefield Publishing Group, Inc.
4501 Forbes Boulevard, Suite 200, Lanham, Maryland 20706
www.rowman.com

86-90 Paul Street, London EC2A 4NE

British Library Cataloguing in Publication Information Available

Library of Congress Control Number: 2022933954

ISBN 9781538105030 (cloth : alk. paper) | ISBN 9781538105047 (electronic)

To the generation who will determine our energy future

Contents

Prologue

Chinese ships of the early 1400s were the marvel of each foreign nation they visited. Using the energy of the wind, the great navy of Zheng He demonstrated the potential for the sailing ship to increase exponentially the nation's contact with the outside world and what would eventually be called its sphere of influence. The technology to harness this source of energy was an essential element for defining this era from the human past; however, at this critical juncture, the nation's cultural will to apply the technology proved to be of even more importance.

Over the last few years, my students in China have told me of the great esteem held for Zheng He in their history. In a nation that today only selectively celebrates its own past, they have shown me that the great sailor has a special—almost universal—importance. But as we proceed to trace the story of energy in world history in our course, they must listen as I present historians' perception of China's great failure to maintain the national will to use the sailing ship to reach outward. In short, China in the Ming Dynasty (1368–1644) introduced the world to the Age of Sail and then stepped aside to watch other nations seize the will to put the technology to use in order to explore and expand. The ensuing era of exploration flows into industrialization, and China is surpassed as one of the world's most developed and technologically advanced nations in what global historians refer to as the "Great Reversal."

In my teaching of energy history to Chinese students, this national choice (while explainable due to the awakening of a cultural and philosophical focus on exclusively providing for its expanding population) marks a point of national failure; a transitional point that determines centuries of human history and national economic stature on a global scale. And it is this same emphasis on energy that provides the opposing view related to humans' current transition from fossil fuels: these same students express great pride in China's pioneering efforts to seize and lead current efforts to innovate human life away from its reliance on fossil fuels. From the shame of missed opportunities of the past, their exuberance of these current efforts suggests a compelling confidence that their nation will have a very different role when future generations study energy history of the early twenty-first century. My work with students at Renmin University in China over a few terms has helped me to organize the approach to energy history that shapes these pages, and I thank them for their assistance.

While each moment in history stands out as unique in its own right, the second decade of the twenty-first century—the moment in which we each sit—is *certainly* distinct. "We have never known so much," volunteered one student at Renmin University, "that we can use to immediately impact our situation." There is great hope as, all around us, new ideas inform our basic understandings of the human condition and how we each live in the world. So extensive are these new perspectives that they force us to reconfigure our own past. It is this need to rethink and organize our history that is behind *To Have and Have Not (THHN)*.

Tracing the history of energy exchange is complicated by the resource and method being put to use. Using the heat of the Sun to grow plants that bear fruit or throwing a log on a fire to create warmth each presents simple transactions that are fairly clear to trace. In more modern times, though, the human experience with energy has possessed a multiplying effect that quickly morphs from a simple exchange to complex cultural characteristics—even an entire way of life. Despite these complexities, crafting the logic of histories such as humans' use of energy allows us to inform and possibly to shape the human future. As humans' most intrinsic exchange with the natural environment, energy holds the key. "Modern civilization is the product of an energy binge," wrote historian Alfred Crosby before continuing ominously: "Binges often end in hangovers."[1] Properly organizing our past marks a promising elixir to such a malady.

NOTES

1. Alfred Crosby, *Children of the Sun* (New York: Norton, 2006), p. 5.

Introduction

The Urgency of Our Relationship with Energy

Cracking ice punctures the heavy, expansive quiet of Antarctica. In 2020–2021, many viewers have become familiar with the cataclysmic sounds of calving glaciers produced by climate change. Earth-shaking fractures splash below to create swells and waves, all of which seem reminiscent of long-ago epochs when continents moved against each other on Earth's surface. The cracking ice on this day, however, is a consistent popping and pressing underlined by the hum of a diesel engine. The source of the sound is the razor-sharp hull of the bright-red icebreaker ship that slowly presses through the icy surface, appearing utterly juxtaposed against the endless surrounding whiteness. Atop the wheelhouse of the icebreaker *M/V Xue Long 2* flies the flag of China, a nation that is thousands of miles distant.

In a moment when a global pandemic has slowed human activity in many normally bustling locales—ranging from New York's Time Square to the newly docile streets of Mumbai, India—activities have increased at each of Earth's poles: Arctic and Antarctic. Particularly at our planet's southern tip, though, sounds such as the Chinese icebreaker have become more frequent. These powerful ships have long been used for fishing in cold terrain; now explorers use them as part of the expanding process of settlement in the form of scientific research stations. Similar to the national flag placed on the unclaimed moon by the United States in the 1960s, such stations stake a claim to this locale and all that is here—or, more importantly, what might be found here. Even a time of global panic can't dissuade the march of progress into new, icy frontiers.

British explorers built one of the region's first permanent structures in 1902 and insulated its walls with the only materials available—felt wrapped around wood. The famed polar explorer Ernest Shackleton recalled that the British station was so drafty and cold that the team remained on the ship and the station became so buried in snow that it was accessible only through a tunnel dug to a single window.[1] Today, by contrast, many of the sixty-eight bases in Antarctica exhibit zippy, prefabricated architecture powered by cutting-edge alternative energy technology that allow them to function as a symbol for each of the seven nations that have laid claim to parts of this region. Without a permanent human population beyond these stations in Earth's known history, Antarctica is a polar desert that has largely been unchanged—ever.

So why, then, have Britain, France, Norway, Australia, New Zealand, Chile, and Argentina drawn lines on Antarctica's map to carve up the empty ice with territorial claims? And why are occupants such as the Chinese icebreaker nosing around to extend those boundaries more in 2022 than ever before? Are they ensuring future access to krill population? Tracing the rate of melt due to climate change? The answer is clear: yes and yes, but not really. Instead, national interests are pinned to the potential mineral and energy supplies becoming accessible due either to melting or the application of new technologies.

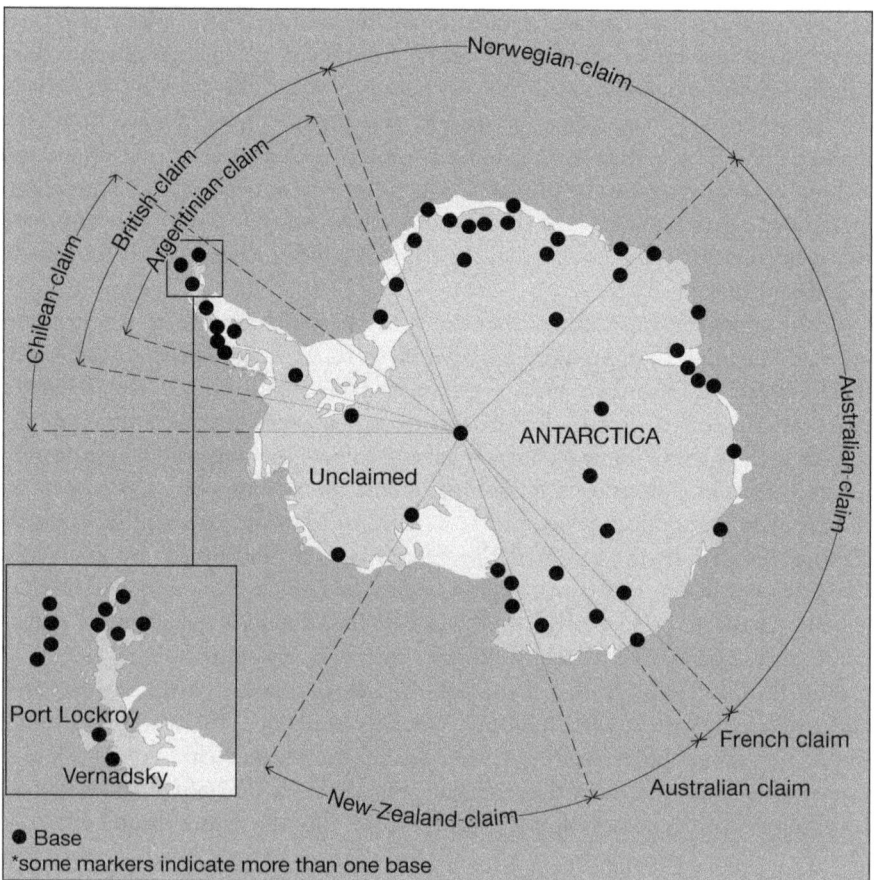

A map of Antarctica territories

Antarctica is not a country: it has no government and no indigenous population. Instead, the entire continent is set aside as a scientific preserve. The year 2019 marked the sixtieth anniversary of the Antarctic Treaty, which was signed by twelve nations in order to ensure that in the region "no new claim, or enlargement of an existing claim, to territorial sovereignty in Antarctica shall be asserted." When the agreement was signed, the region was primarily not to be used for military purposes. Scientific research was permissible and has obviously offered nations the most viable fashion in which to ensure a place at the table, should Antarctica ever become relevant; however, scant other importance could be imagined in the 1960s. But new realities have emerged and other member nations have watched warily since 2015 as China—with its multiple icebreakers and four research stations (Great Wall, Zhongshan, Taishan Summer Camp, and Kunlun/Dome A)—has markedly grown its presence in these icy stretches.[2]

To complicate matters for this sovereign region, over the last few decades scientific research—specifically, geomorphological mapping—has indicated a bounty of energy resources beneath Antarctica, this unowned continent: estimates have discussed reserves of two hundred billion barrels of oil, far more than in some oil-rich areas, including Kuwait and Abu Dhabi. With the protocol banning mineral prospecting set to expire in 2048, the icebreakers have come here and interest has intensified. In March 2020, for instance, Russia's state-run geological survey firm, Rosgeologia, reportedly undertook a major new seismic survey in the Riiser-Larsen Sea, off the coast of Antarctica's Queen Maud Land. Moreover, Rosgeologia stated unequivocally that it did this 4,400-kilometer survey—the first seismic survey done in the area by Russia since the late 1990s—with the express purpose of "assessing the offshore oil and gas potential of the area using the latest technology."[3]

Throughout history, global power required that nations ensure supplies of their most essential commodities. Activity in Antarctica is just one example of how energy has come to top this list for many nations in 2021.[4] After centuries of alterations to human living patterns, abundant and reliable sources of power now are indispensable. The unseen hand that enables our cell phones, automobiles, blenders, and air conditioners, energy appears to many of us to be as essential as air, food, and water. Without such resources, it seems to most of us, humans can't live. Therefore, these nations look to the undeveloped resources of Antarctica with a wary eye toward global competitors. It is just one example today of the radical shifts occurring in how humans in general see their place in the world.

In one of our era's most unique developments, across disciplines ranging from geology to philosophy, a growing number of scholars now refer to our imperfect moment as the Anthropocene Epoch: a moment marked by Earth being perilously near to various cataclysmic tipping points—largely due to the discernible causality of human activities such as burning fossil fuels—and yet recognizably salvageable, with the help of human ingenuity and insight. Indeed, the concept presents a remarkable meeting of various "economies of knowledge" that force diverse nations to find a common ground. In terms of intellectual history, the Anthropocene concept in this vein marks a truly culminating point of our understanding of humans' place in nature.[5] We have moved from lone calls—such as George Perkins Marsh in 1864 that "man is everywhere a disturbing agent" and Rachel Carson in 1961 saying that our chemicals and practices have impacts throughout the ecological webs of life that surround us—to achieve a scale and scope today that is global and even geological in our grasp of humans' impact on Earth.[6] Currently, our scientific knowledge allows us to discern the scale of the global disruption wrought by the remarkable advancement of our species.

Just as the Anthropocene portends a common forecast for the future of all humans, it also affords a way of unifying our past. In its ability to encompass broad, diverse elements of the human condition and also to allow historians and others an opportunity, at last, to structure a framework to the human past that more accurately contextualizes our species within behavioral details that overcome political and ethnic borders, the Anthropocene concept reorders our worldview. In addition, more than any previous intellectual shift, the Anthropocene concept provides us with an imperative to tell a better, more complete story of our past. Primarily, historians now have the opportunity to apply macroscopic organizational concepts to human history in order to better overcome global borders and to bring to the fore elements of global ecology that are held in common.

United in the human future, historians can help to weave our understandings to better reflect a common past—a history that, while making room for the experiences of distinct cultures and societies, also reveals the emergence of patterns that sync humanity together. For its survival, our species relies on Earth's basic elements, including air and water, and its day-to-day living relies on exchanges of environmental elements that—at their foundational level—each involve the release or harvest of energy. As theorist Vaclav Smil writes, "Energy is the only universal currency; one of its many forms must be transformed to get anything done."[7] For the Anthropocene to gain coherence, our histories must amplify and emphasize such common human traits.

At the root of the epoch that defined the twentieth century was a clarifying new aggressive fashion of viewing energy: flexible sources in the form of fossil fuels that might be gathered in one locale, assigned a value, and then used elsewhere. Important new controls grew from the global experience with energy in the 1970s and transformed a complex new stratification of global power from either a nation's need for or its possession of energy resources. Finally, just as new ideas have formed the Anthropocene concept at the start of the twenty-first century, the view of energy has also been transformed with a clear eye toward sustainability and long-term growth while minimizing harm to the environment.

Witnessed by the growing activity in Antarctica, such innovations now appear as a crucial facet of national security—both to ensure supply and to minimize its necessity.

Throughout the Anthropocene, energy has remained a vital cog in the lives of modern humans. *THHN* discusses many such moments that represent significant leaps in our conceptions of energy; however, historians have not yet knitted this story of exchange together in order to trace its overall continuities. The Anthropocene concept demands that we organize our energy past not only for physical technological transfers but also for exchanges of ideas—such as sailing technology and how it plays out differently in societies ranging from China to Portugal. Indeed, the development of the economic engine of the modern world begins with the transition to wind energy and then derives from ancillary ideas, ranging from navigation and geography to theories such as capitalism. Such a broad revolution in thinking began with a basic reconfiguring of our concept of the human's place in nature based on new discoveries and understandings. Instead of pulling back from such dynamic moments, the most successful nations seized the revolution and synched their futures to it.

Similarly, in 2022 climate change might be seen as a paradigmatic alteration in our place on Earth based in new science and technology. In short, computer modeling and data collection from carbon dating and satellite views reconfigured our knowledge of Earth's climate. New information informed a clearer view of the impact human development had on Earth, particularly in the era of the massive burning of fossil fuels for energy. It is these lessons of climate change, for instance, that drive China's icebreaker (as well as those of the six other nations) into the Antarctic region, not only to collect scientific data but also to situate each nation for what dwindling opportunities remain of the era of fossil fuels. At the start of the twenty-first century, the context provided by our new understanding of Earth's past and future seems to have proven even more revolutionary than the realization in the 1500s that our planet was round.

It is imperative that this history informs the gravity with which we approach our energy transition today as we apply new knowledge such as climate change. *THHN* periodizes transitions from one energy source to another and also creates demarcations that properly acknowledge more significant shifts to provide a common language that allows us to catalogue the ongoing human relationship with energy. Formalizing such distinctions sharpens our understanding of the differences between energy sources and user societies. In particular, as we study and organize specific energy transitions, we must differentiate among changes in similar sources so that we might distinguish those that are more consequential—what we will call *intrinsic* or *foundational* shifts in energy use—in which the ethic or mode of collection or use marks a severe departure from previous models. Defining this distinction is particularly crucial as we make energy studies a vital component of efforts to organize the Anthropocene concept. In order to properly contextualize the changes in human living during this geological epoch, it is particularly imperative that we distinguish between the transitions in the ethic or mode of harvest. To accomplish this task, we utilize principles of physics, which, of course, studies energy and matter in nature. In *Energy and Civilization in History*, Smil suggests such a foundation in physics by offering language to help historians distinguish between sources of energy, including:

- **Efficiency of energy conversions**: the effectiveness of technologies employed to harvest or convert raw materials into energy
- **Energy returns**: which need to exceed input for a source to be viable and sustainable
- **Energy intensity**: which measures the cost of products, services, and even of aggregate economic output—however, it can't necessarily include past costs or inputs. Thus, such calculations are not necessarily cumulative.[8]

With the implications of the Anthropocene in mind, energy transitions of the nineteenth century are most acute as we trace the uneven adoption of the fossil fuels that will mightily increase the impact of humans on Earth's environment. In *Fossil Capital*, historian Andreas Malm creates a template for connecting these new sources of power to their future impacts when he deems burning of coal, petroleum, and other fossil fuels after the 1700s as "so many invisible missiles aimed at the future."[9] Particularly at critical junctures in our energy past such as the mid-nineteenth century, transitions merit particular importance and should be subject to careful scrutiny. Informed by the ideas of Smil, *THHN* subjects sources of energy to questions, including:

- What is required to harvest this energy source?
- Is there an energy investment in its harvest? If so, how does it compare to its output?
- Are there basic ways in which the source is inherently limited in its accessibility, supply, or scope?
- Can the energy extracted from this source be applied to other applications as well?
- What are the source's capabilities for expansion and further development?
- Are there significantly different outcomes or waste from this source as opposed to others?
- Regarding transitions, how does this source of power amend or replace another?
- Regarding transitions, in its collection or use, does it represent a significant change from previous sources?

The answers to such a line of questioning distinguish each energy source from another and, in so doing, also emphasize certain transitions as more critical than others. While each transition is important, the *modular* shift reflects much more dramatic implications for human users. In distinguishing the fossil fuel era from those that preceded it, for instance, Smil writes:

> The contrast is clear. Preindustrial societies tapped virtually instantaneous solar energy flows, converting only a negligible fraction of practically inexhaustible radiation income. Modern civilization depends on extracting prodigious energy stores, depleting finite fossil fuel deposits that cannot be replenished on time scales order of magnitude longer than the existence of our species.[10]

The clear lesson of *THHN* is that at moments such as the emergence of the Age of Sail in the 1500s, the nations that seized the new logic and found ways to augment and supplement—not to resist—a new worldview found unexpected opportunity during the ensuing century. Similarly, the expansion of the use of fossil fuels in the twentieth century catapulted the global standing of nations with the required supplies and technologies. Now, new realities and knowledge force our species to flex in promising directions. History will note the nations that seize and lead a movement toward more sustainable modes of power and those that fail.

A NOTE ON ORGANIZATION

Each chapter of *THHN* begins with a GROUND element that anchors the main idea, similar to how a ground wire assures and stabilizes the flow of electricity. Often, the GROUND is a specific locale, experience, or item. Organized around an overall pattern of use or awareness, each chapter relies on an assortment of global vignettes that contribute to our species' relationship with energy.

Additionally, within each chapter, I utilize two elements to amplify certain patterns or trends that I will define for readers here:

CONDUIT: The technical stuff of harvesting, making, and managing power
CURRENT: The ideas that make energy power

NOTES

1. Simon Watkins, "Russia Makes Move on Antarctica's 513 Billion Barrels of Oil," oilprice.com. https://oilprice.com/Energy/Crude-Oil/Russia-Makes-Move-On-Antarcti cas-513-Billion-Barrels-Of-Oil.html. Accessed January 6, 2022.

2. Nengye Liu, "What Are China's Intentions in Antarctica?" https://thediplomat.com/2019/06/what-are-chinas-intentions-in-antarctica/. Accessed January 6, 2022.

3. Watkins, "Russia Makes Move."

4. Leah Feiger and Mara Wilson, "The Countries Taking Advantage of Antarctica During the Pandemic." https://www.theatlantic.com/politics/archive/2020/05/antarctica-great-power-competition-australia-united-states-britain-russia-china-arctic/611674/. Accessed January 6, 2022.

5. The reference here is to the discussion by natural scientists and humanists to delineate our current geological epoch as the "Anthropocene" because humans have created significant impacts on the environment and they control Earth's fate. See, for instance, John R. McNeill, *Something New Under the Sun: An Environmental History of the Twentieth-Century World* (New York: Norton, 2001).

6. The evolution of much of this environmental thought, however, required changes to our standard of living, particularly in the United States. Ironically, one could argue we wouldn't have arrived at this realization without a century of humans living the "high energy life."

7. Vaclav Smil, *Energy and Civilization: A History* (Cambridge: MIT Press, 2017), pp. 12–14.

8. Smil, *Energy and Civilization*.

9. Andreas Malm, *Fossil Capital* (London: Verso, 2016), pp. 7–9.

10. Smil, *Energy and Civilization*, p. 295.

I

ENERGY EXCHANGE IN THE BIOLOGICAL OLD REGIME (BEFORE 1400)

Animal power intensified the capabilities of human workers. LIBRARY OF CONGRESS PRINTS AND PHOTOGRAPHS DIVISION

Animal power, harnessed for agriculture in this image, compounded the ability of humans to grow food through photosynthesis in an extension of energy patterns begun during the biological old regime.

As a system of energy exchanges, the natural environment hosts each species and provides what is needed for it to satisfy basic needs—to survive. Although our more modern history distinguishes us from other species, humans' distant past associates us with all other living things. Over time, our story and our existence has become complicated; however, the point of origin is undeniable and simple. Humans first strove for survival just like all other living things, and it was unthinkable that we or any other member of Earth's system could disrupt the scale and scope of its innate patterns.

1

Energy in the Human Past

GROUND: AMSTERDAM, HOLLAND

The paddles swing grandly outward thirty feet and as the breeze gently pushes they swing clockwise, first slowly before using gravity and weight to push downward with growing force. Just as the paddle reaches its bottom rotation, the process repeats as the genius of a second, third, and fourth paddle repeat the effort to capture moving air. With only a slight breeze, the windmill can continue to spin while metal gears further the magic by transferring the motion to milling implements hidden behind walls behind the turbine. Together, the windmill efficiently transforms the wind's energy to grinding stones that turn grain into flour. In Amsterdam, Holland, though, such a mill serves as part of a complex wind energy landscape that transformed the globe.

Although windmills were often found in the Dutch countryside, in Amsterdam by the 1500s they could also be found along the city's system of man-made canals. Boats move through the canals by using a variety of power sources; however, the canals were built as thoroughfares for traders who used wind power to travel all over the globe. Admirable plumbers, Amsterdammers (similar to other humans living in submerged landscapes) very early on learned how to dry out swampland in order to convert it for habitation and agriculture. In their case, they constructed systems of dikes and dams to keep out the sea while also slicing channels through the peat bogs. "The water, the perils, the bravery, the absurdity of the geographic position, and the development of complex communal organizations to cope with the situation," writes historian Richard Shorto, explains Amsterdam's "never-ending struggle against nature."[1] From this common technological start with a dam along the Amstel River sometime after 1200, though, the Dutch distinguished themselves from other cultures.

As early trading ships plied regional oceans, the idea of a port occurred almost organically along well-traveled routes. While much of medieval Europe organized

11

around the manorial, top-down society (whether owned by a church or private citizen), the Dutch province's lands were not locked up by a noble class.[2] This independence gave the Dutch flexibility to grasp new opportunities that emerged. Combining this realization of larger trade systems with early engineering ideas, starting in the 1300s Amsterdam used the world's most complex and enduring system of man-made canals to create a hub of privately owned warehouses (known as canal houses) that provided early traders both residence and storage. These planned circular canals served as thoroughfares that stratified economic standing for the traders while also providing needed access between open sea and storage.

Before the Age of Sail had even truly been defined, Amsterdam stood as its version of the modern-day Amazon.com: distribution and storage hub for goods and services being ordered and purchased throughout the globe in an increasingly complex web of a trade system. The system became more official in 1599 with the establishment in Amsterdam of the Dutch East India Company (Vereenigde Oostindische Compagnie; VOC). "It's by no means a stretch to say that the VOC remade the world," writes Shorto. "And to a large extent Amsterdam made the VOC."[3]

With intentionality, Amsterdammers applied scattered innovations of the biological old regime and organized a city around them. They were engineers when there was little technical knowledge to be applied. By accepting the need for humans to gather in central locations, fearlessly working with and understanding their local environment, and perceptively applying ingenuity as it emerged, residents of this early trading port poised the city to define the earliest period of humans' shift toward industry.

Prehistory is a term based in our modern bias. It assumes that our history begins when humans live in a certain, prescribed fashion. Additionally, it assumes that anything that occurred prior to this point was less important or consequential. One of the strengths of using energy to organize humans' chronology is that we can restore certain continuities to our story—much like the perspective of human geography or environmental history. When viewed as a species existing in its natural environment, our early history reveals humans' core functions and needs. Regardless of the biome in which humans resided, their survival depended on natural exchanges that captured or released energy for various purposes. There was no human existence without such definition—no story for us prior to that.

As a system of energy exchanges, the natural environment hosts each species and provides what is needed for it to satisfy basic needs—to survive. Although our more modern history distinguishes us from other species, our distant past associates us with all else. The level of treatment in the pages of *THHN* is primarily predicated on how impactful human living patterns were on Earth. Although the overwhelming majority of our species' history was spent as hunter-gatherers who worked within a pattern of simple energy exchanges, our impact on the environment was slight. The pages of *THHN* will primarily emphasize the impactful exchanges of the postindustrial human. In short, today, our story has become complicated; however, the point of origin is undeniable. Humans first strove for survival just like all other living things, and it was unthinkable that we or any other member of Earth's system could disrupt the scale and scope of the planet's innate patterns.

As its point of origin, Earth's energy patterns begin with the Sun that radiates from the center of our planet's universe. Earth is mere debris left over from the Sun's formation, and more than one million of our planets would fit in it. The Sun exists as the center of the life on which depend humans and all the living things that we require. In short, its energy makes possible our very existence.

From this beginning, changes in human living can be traced through methods of acquiring and mastering energy. More than any other portion of the human experience, our interaction with energy fits snuggly in the paradigm with which Jared Diamond organizes all of our story when he writes: "Much of human history has consisted of unequal conflicts between the haves and the have-nots."[4] Having energy, of course, is only the beginning. Societies have also needed to choose what to do with it or how much of a cost it warranted. And living with new regimes of energy has fueled societies to develop in dramatically different ways from each other as well as from those without. In Amsterdam, for instance, we will learn of a society that seized energy convergences to help define a new era of human living. As the pages of *THHN* tell this story, however, our effort to organize our energy past requires that we begin with the Sun.

"Today, as ever," writes historian Alfred Crosby of humans' use of fossil fuels, "we couldn't be more creatures of the sun if we went about with solar panels on our backs."[5] In fact, though, compared to the past, the connection of today's energy patterns has become quite abstracted from the Sun from which all energy originates. Past humans' first energy interactions were drawn much more directly from the Sun's heat.

Throughout our existence, the two principle sources of energy for humans have been upwellings of Earth's heat, such as magma, and radiation from the Sun. Physicists have quantified and measured each of these sources of power, and humans have innovated to extend the systems and capabilities of these original sources. The systems are often so complicated that contemporary humans living in a world made possible by pipelines, powerlines, and transformers can often lose track that they remain "children of the sun."[6]

Life began on Earth millions of years ago at the level of bacteria, particularly cyanobacteria, or blue-green algae. The growth of such organisms relies on sunlight in the interaction that we now refer to as photosynthesis, which is the basis for plant and animal life as it moves through the food chain. In the process of photosynthesis the oxygen component of the broken water molecules drifts away and becomes most of the oxygen in our atmosphere. It is also the oxygen that enables human respiration—breathing—that is required to drive muscles and, therefore, to do work.[7]

Some of humans' first "work" was the basic activity that enables our species to survive: hunting and gathering. In fact, 99 percent of our species' history has been spent as hunter-gatherers. Throughout various ecosystems, humans' success—our ability to survive—derived first and foremost from our basic energy exchange, drawing a breath of oxygen. What we sought to accomplish with our sustaining breath, then, was typically to collect what we needed to sustain our bodies, and each of the quarries of our hunting and gathering grew in one way or another from an exchange of the Sun's heat. From the first efforts to perform these tasks of survival, practitioners sought to perfect or systematize their tasks. As routines and systems evolved, humans all over the globe achieved varying levels of effectiveness and ultimately reached the level of domesticating the activities—making them part of their species' patterns of everyday life.

Patterns of movement marked one of humans' first great departures: bipedalism. Upright walking liberated the hominin's arms for doing other things, such as carrying weapons or collecting food and taking it to group sites for later consumption. This method for moving through the landscape, then, over time, allowed for other developments, particularly the adoption of wooden tools (sticks and clubs) and those also incorporating stones. Regardless, any such device could not alter the basic reality of the hunter-gatherer: they could only rely on their muscles and simple cognitive analysis to secure food. Even the tools of the era merely extended, focused, or adapted human energy.

While still relying largely on human labor, agricultural advancements occurred through control and management of crops that occurred largely with experience that would then be passed along through societies. Diamond stresses that for hunter-gatherers, such advancements didn't necessarily mean less work. "In reality," he writes, "only for today's affluent First World citizens, who don't actually do the work of raising food themselves, does food production mean less physical work, more comfort, freedom from starvation, and a longer expected lifetime."[8] From our modern perspective, the deprivation of living as a hunter-gatherer seems so obvious; yet, as Diamond explains:

> In each area of the globe the first people who adopted food production could obviously not have been making a conscious choice or consciously striving toward farming as a goal, because they had never seen farming and had no way of knowing what it would be like. Instead . . . , food production *evolved* as a by-product of decisions made without awareness of their consequences.[9]

Therefore, the best view of human existence in the biological old regime is that food production proceeded independently in only a few areas of the world in which knowledge accumulated through decisions over allocating time and effort, and success came much more due to "differences in continental environment, not in human biology."[10] In these areas, reliable food production allowed societies a head start on the path of development that proceeded through technology, political organization, and other features of complex societies that could emerge more readily in dense, sedentary populations with sufficient food to survive. "The result," writes Diamond, "was a long series of collisions between the haves and the have-nots of history."[11]

Many of the "collisions" (maybe *encounters* is a more representative term) that ensued occurred among the societies that gradually began to reach outward, on land and sea. The earliest exchanges occurred via human movement and, steadily, due to the increasing abilities of seafarers to venture farther from their home region. Muscle power as a prime mover was not sufficient to cross more vast oceans. Mastery of emerging energy systems was required to move beyond the small-scale travel of the Mediterranean and Polynesian worlds.[12]

Thus, prior to these interactions, each human lived within its domain by using essentially the same patterns of interacting with the environment and with only minimal energy exchange. During this period of the biological old regime, the nine pockets of evolving human agriculturalists pursued distinct tracks as society. Certain similarities might be established; however, one basic theme of human life through roughly 1400 remains the lack of a global, coherent pattern for human advancements. Instead, it becomes remarkable how humans with little to no contact with others pursued similar but distinct efforts to survive and, ultimately, to thrive. From common origins and knowledge, they worked within the limited nature of the biological old regime while clearly reaching forward to define a collective future that would be starkly different for many humans.

CALCULATING THE NET ENERGY COST OF HUMAN LABOR

Early humans operated within the basic constraints of personal survival. Foraging, or the search for foodstuff, is the basic characteristic of human survival. The body requires energy to perform the activity, and this is largely based in nutrition as we eat in order to replenish the body's capabilities. Hunting merely added the complication of pursuing animals or fish for meat, yet it added considerable capacity for complicating things. Hunting introduced variables to human existence ranging from weapons and tools to patterns of movement. Almost entirely defined by climate and species, patterns of hunting might at times resemble collection (such as nuts and mollusks) to confrontation with animals that required more advanced tools. Often done in groups, hunting animals often involved pursuing injured animals as well as butchering and transporting the meat. Organized, communal hunting brought the greatest reward, and different societies practiced methods for herding or steering animals over cliffs or into pens. In such a way, many large herbivores (such as mammoths, bison, deer, antelope, and mountain sheep) could be slaughtered to provide a sufficient supply of meat to support an entire community.

A UNESCO World Heritage Site near Fort Macleod, Alberta, provides evidence of the favorable energy exchange that went into such directed hunts for over five thousand years. Late Pleistocene hunters steered herds of buffalo into drive lanes before more directly scaring the herd over the cliff.[13] It is easy to see how the addition of horses would have only made such techniques more effective. Particularly in such examples of mass killing, hunting, more importantly, compounded the human interest in food preservation and processing that stimulated the use of cooking. Cooking marks one of the first significant shifts in humans' use of energy.

Crosby clearly states: "Cooking is cultural, not genetic, an unprecedented innovation."[14] And yet cooking has been found to be universal among our species and is more unequivocally characteristic of our species than language.[15] Archaeologists believe that *Homo erectus*, who migrated out of Africa and across Eurasia approximately 1.7 million to 200,000 BP, used fire as an aid for food preparation and management.[16] Most often, women gathered kindling and biomass for burning and tended the fires. This process of moving from being awed by fire and its natural occurrence to managing and planning for its use is referred to as domestication. Begun by collecting the coal remnants from natural fire, humans eventually rubbed sticks together or with flint stone to create a spark. Hearths begin to appear in settlements by forty or fifty thousand years ago, and archaeologists use the prolific appearance of them as one of the pivot points to define the Upper Paleolithic period. This revolution in energy use defines a clear transition in hunter-gatherers that, then, ultimately defines a basic dimension of the revolution to sedentary agriculture.

In Upper Paleolithic sites, archaeologists have found high-grade flint, amber, and seashells far from the objects' points of origin. Most often, cooking took place by suspending meat above the flames, buried in hot embers, placed on hot rocks, encased in a tough skin, covered by clay, or put with hot stones into leather pouches filled with water. Cultural development also seems to have accelerated at this point with a growing prevalence of art and the creation of more permanent shelters in caves or crude houses. In areas such as Siberia, we also find the practice of burying valuable objects in graves begin in earnest. Finally, in the cave paintings of Franco-Cantabria from the Upper Paleolithic period one finds record of wild animals that were long gone from the region, including bison, horses, reindeer, and mammoths. These innovations fueled the development of speech and patterns of language that also grew out of settled living.

WOOD CONSUMPTION IN OPEN
FIRE COOKING OF MEAT

The heat given off by fire primarily served important roles for early humans as a source for warming and cooking; however, it also functioned as an early tool, separate from the heat it produced. Strategically, early humans applied fire to purposes such as treating stones so that they would better flake or split and as a tool for environmental management. Applying fire for landscape manipulation could serve as a hunting tool (either steering populations or clearing brush to improve movement or visibility).

These cornerstones of our species' use of energy provided the raw material for millennia of living leading through a clear demarcation when the increase of the domestication of each of these activities allowed for an increasing level of control. Extending beyond their needs for mere survival, humans unevenly moved from life as hunter-gathers to variations of a sedentary life. As they did so, however, energy became more important, not less.

Domestication of these early techniques and technologies took many forms; the primary outcome, though, was an overall organization of living patterns that is often referred to as the agricultural revolution. In short, this revolution involved each of the uses of energy realized by hunter-gatherers in small-scale or diffused examples. By evolving these methods for living, humans became agriculturalists and organized much more reliable, steady production of food crops by harvesting, accommodating, and managing the energy of the Sun. As Crosby points out, "The hunter-gatherer way of exploiting sunlight had been successful," but for whatever reason humans wanted more stability and reliability. He writes:

> We know as an archeological certainty that humans in many regions intensified their exploitation of plants and smaller animals, learning not only how to harvest both, but how to encourage their propagation and availability for future meals.[17]

By approximately 4000 BP humans had domesticated almost every crop plant and animal essential to today's civilization, including wheat, rice, barley, potatoes, dogs, horses, cattle, sheep, and chickens.

Through domestication, humans married plants and animals with the physical structure and composition, life cycles, and other details of region and climate. It occurred all over the globe at different rates. Often, plants and animals such as horses would be harnessed so that their energy served human purposes and needs. Taken in general, livestock were the perfect vehicle for marrying the energy cycles of nature to human needs and outcomes. A farmer could grow grain and use it to fill livestock for the purposes of providing food or work or both.

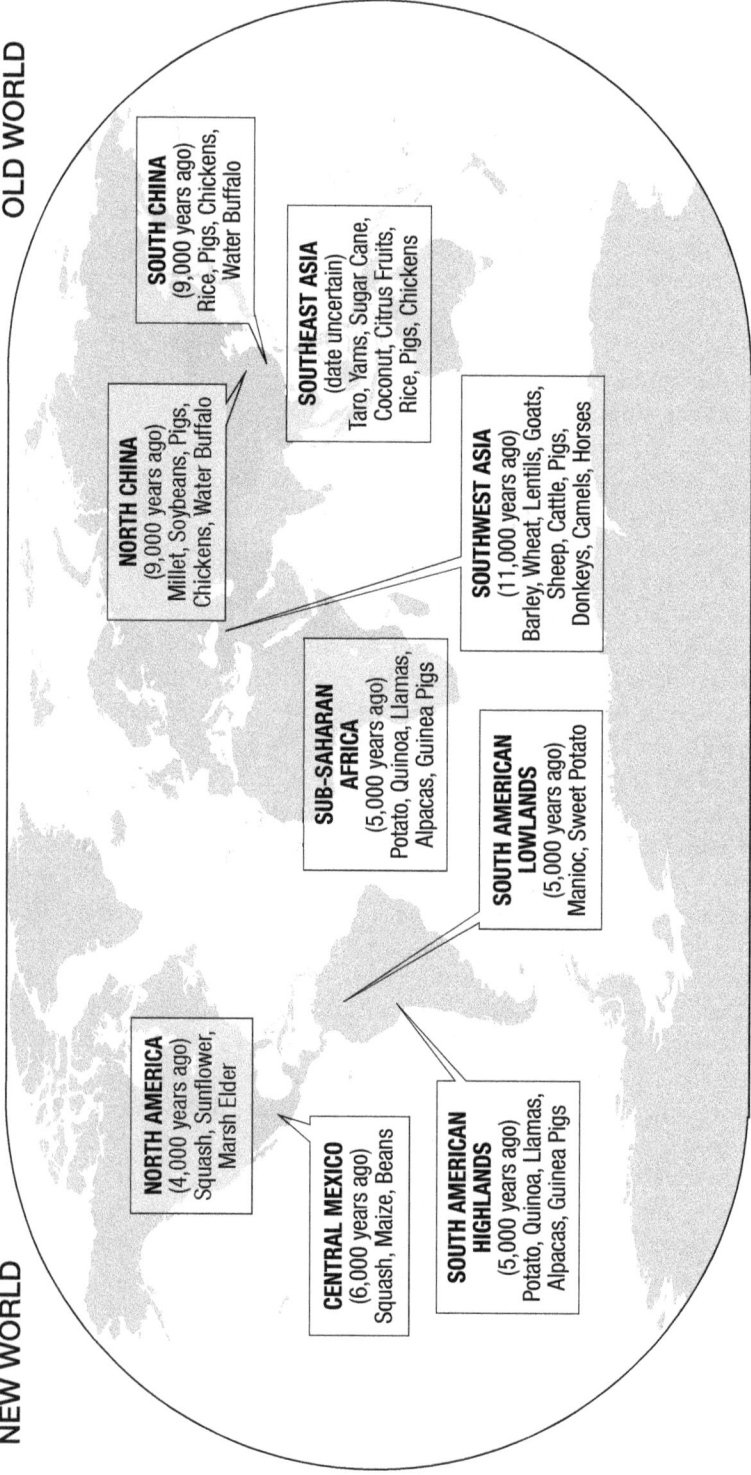

NEW WORLD

OLD WORLD

SOUTH CHINA
(9,000 years ago)
Rice, Pigs, Chickens,
Water Buffalo

SOUTHEAST ASIA
(date uncertain)
Taro, Yams, Sugar Cane,
Coconut, Citrus Fruits,
Rice, Pigs, Chickens

NORTH CHINA
(9,000 years ago)
Millet, Soybeans, Pigs,
Chickens, Water Buffalo

SOUTHWEST ASIA
(11,000 years ago)
Barley, Wheat, Lentils, Goats,
Sheep, Cattle, Pigs,
Donkeys, Camels, Horses

**SUB-SAHARAN
AFRICA**
(5,000 years ago)
Potato, Quinoa, Llamas,
Alpacas, Guinea Pigs

**SOUTH AMERICAN
LOWLANDS**
(5,000 years ago)
Manioc, Sweet Potato

NORTH AMERICA
(4,000 years ago)
Squash, Sunflower,
Marsh Elder

CENTRAL MEXICO
(6,000 years ago)
Squash, Maize, Beans

**SOUTH AMERICAN
HIGHLANDS**
(5,000 years ago)
Potato, Quinoa, Llamas,
Alpacas, Guinea Pigs

Timeframe of when plants and animals became domesticated.

Simply, it was a more efficient exploitation of solar energy than hunting and gathering that enabled people to live on less land and remain in one spot.[18]

During this early period, *Homo sapiens* perfected agricultural patterns in nine separate regions: Southwest Asia, Southeast Asia, North China, South China, and sub-Saharan Africa in the Old World; and eastern North America, central Mexico, the South American highlands, and the South American lowlands in the New World. In accordance with their climate strengths and weaknesses, each of these centers domesticated its own species of plants and animals. For instance, maize, squash, and beans became prevalent in central Mexico and livestock did not. By contrast, South China focused on rice and also pigs and chickens. The primary tool that was employed to harvest and manage the Sun's energy was the human body. Very little sharing occurred between the Old and New Worlds until the second millennium CE.

NEW TECHNOLOGIES EXPAND ENERGY POSSIBILITIES

Humans by 1200 had clearly begun to investigate methods for applying crude forms of technology to overcome the limits of their own bodies' ability to work. Transitioning to the use of such innovations steadily graduated, as well, to organized examples of proto-industries where various types of communities could band together and follow a plan for unified development. That said, although domestication remained a primary dimension of human society so, too, did the limitations of existing technology, methods, and energy. Across a wide spectrum of examples and regions, humans repeatedly butted up against the restrictive nature of their tools to do work. While innovation might extend capabilities of work, the basic limiter remained those of the animate sources of energy that remained the most important prime movers until the middle of the twentieth century. Societies that derived their kinetic energy almost solely or largely from animate power, writes Smil, "could not provide a reliable food supply and material affluence to most of their inhabitants."[19] Increasing the strength of these sources of power was accomplished by either concentrating individual inputs or by using mechanical devices to redirect or amplify muscular exertions.

Remaining, by definition, in the biological old regime, these humans worked within the confines of their agricultural capabilities. In general, the routes toward intensification focused on three essential but erratic and vacillating advances: replacing as much human labor as possible with that of animals; implementing irrigation and fertilization to overcome limits of climate; and growing a greater variety of crops, using rotations or multicropping techniques. Beyond simple conversion of the Sun's energy to plants and food, during the biological old regime, certain innovations that appeared, in hindsight, to be capable of life-altering possibilities were denied "takeoff"—in the words of historians of technology.[20] Overarching political or economic logic, overriding concerns with survival, or—most particularly—restrictions on energy input undercut the scale and scope of what could be done by humans prior to 1400. To enable takeoff, food supplies needed to become stable and markets needed to form that could create economic mechanisms to support infrastructural growth.

CONDUIT: Extending Human Power through Leverage

During antiquity, a variety of societies increased the abilities of humans by employing levers, which were typically rigid, slender pieces of wood or metal. Technically, the oar or paddle in a boat is one of the earliest and clearest examples of how this simple innovation might multiply energy and thereby the capacity for work to be completed. However, on land, levers were a basic tool that could be directed to tasks by adjusting the fulcrum point. Of course, the work was still only completed through the exertion of a human body.[21]

Diversification of the lever technology can be seen between devices such as a crowbar and a wheelbarrow. Each item functions as a lever, but

Windlasses and levers were basic technologies that humans could apply to a variety of tasks through the use of hemp ropes or metal chains. LIBRARY OF CONGRESS PRINTS AND PHOTOGRAPHS DIVISION

the wheelbarrow added the technology of the wheel to allow a load to not only be lifted but also moved. Ancient Chinese workers often employed these means and are also credited with expanding the application of pulleys, which built on these other technologies but added rope. These were also some of the essential technologies of the rigging in sailing ships.

Windlasses and capstans, two devices primarily known for their use on ships, had earlier applications on land that applied the simple technology to purposes ranging from lifting water from a well to creating the tension that enabled many early weapon systems, such as catapults. Most important, similar to their purpose onboard ships, windlasses and capstans lifted heavy loads.[22]

Finally, treadwheels demonstrate how early technologies could be flexed to accept power from various prime movers. The paddles of the wheels might act as footing for humans who were stepping to push the wheel around, or animals might be substituted. Finally, very similar wheels would be used to draw power from rivers. Throughout the history of energy use, this type of flexibility facilitated transitions to new power sources.

ANIMAL POWER JOINS WITH HUMANS TO EXPAND AGRICULTURE

Globally, innovations in agriculture proceeded at varying paces. Egyptian farming used irrigation technology to create a cradle of farming that emphasized managing annual floodwaters. In the Indus valley and Mesopotamia, canal irrigation and methods to lift water to higher-lying farms created stable food production despite the dry climate. In China, agricultural methods remained largely rudimentary. By the time Egypt was supplying the Roman Empire with surplus grain (during the Han Dynasty, 206 BCE–220 CE), Chinese farmers adopted methods and tools (including iron moldboard plows, collar harnesses for horses, and seed drills) that Europe and the Americas would not adopt for centuries. The Han Dynasty marked a dynamic period in Chinese agriculture that, afterward, would largely remain stagnant for centuries. This golden era was marked by the development of extensive irrigation systems, including the most famous example, Sichuan's Dujianhyan, which still grows food for millions of Chinese today.

The use of draft animals distinguished these societies from the agricultural development of Mesoamerican civilizations. Corn tied together the civilizations of the Mayan lowlands with the high-lying basins of Mexico. Many archaeologists attribute the massive decline in Mayan population after the eighth century CE to unsustainable agricultural practices (primarily excessive erosion and a breakdown in their complex water management program). A succession of complex cultures developed throughout the basin of Mexico with the largest population relying on permanent canal irrigation around Teotihuacan. Throughout Mesoamerica, agriculturalists relied on the use of chinampas, raised rectangular fields that utilized excavated mud, crop residues, and water weeds. Chinampas were masterful at turning unproductive swamps into productive and sustainable farms as early as 100 BCE and peaked during the last decades of Aztec rule. Irrigated corn was the crop around which these cultures organized themselves.

In Europe, agricultural production was generally inferior to that of China until the seventeenth century. Emerging from the Middle Ages, European agriculture was quite limited and trailed that seen in many other regions. In particular, Muslim agriculture was passed on to their European counterparts through areas such as Spain where Spanish Muslim farmers passed on their knowledge to the Spanish Christians. Of particular significance, Muslims were the first to identify that floral diagrams could be used for purposes for classification, knowledge that provided the base for many Renaissance botanists to advance the studies of plants. In Europe, agriculture became increasingly important after Roman times not because of innovation but more because the chief basis of wealth and of political power was the land. What became known as medieval feudalism was adapted to a state of society in which land was the source of wealth and military force was the basis of power in an agrarian society under siege from neighboring powers. Indeed, after the breakdown of Rome and its institutions, the powers of Western Europe were subject to attack by Muslims, Norsemen, and Hungarians. Until about the end of the thirteenth century, feudalism succeeded fairly well in maintaining order.

Beginning from a foundation in agriculture, social and political structure in Europe emphasized land tenancy of one sort or another. The word *feudalism* derives from the Latin term *fief* (in Latin, *feudum*), which referred generally to a grant of land from one nobleman to another; typically, the granter of the fief was the lord and the recipient was his vassal. Each lord could have several vassals, and vassals could have several lords. However, the ordering of responsibilities was clear: The lord had to protect his vassal; and the vassal, to serve his lord. Within this structure, the basic unit was the manor, which functioned as an agricultural estate that belonged to a lord, who served as a member of the noble class. By and large, most of the manor's inhabitants were peasants, whose basic job was to cultivate the soil for the lord's benefit. The society did not allow them to aspire to owning their own land.

On a typical European manor, the very organized farming was carried out in a three-field system that stimulated a primitive kind of crop rotation. In this system, one part of the land was planted in any given year with a winter crop and another with a spring crop while the third part was allowed to lie fallow, which was considered the best method for preventing soil depletion. The roles of each field would be rotated annually. Within this organization, the peasant on a manor would be given jurisdiction of specific strips in those three fields. Kept separate, the lord's own holdings were referred to as the demesne or domain lands and were typically scattered among the fields. Most often, the lord's block included a solid block nearest his residence. One of the primary control mechanisms was the cost of agricultural tools (including animals), which were owned by the manor and used by individual peasants.

Through the manor system, agriculture was a product and extender of the rigid class system that made up feudalism. At no time, however, did the manor and manorial system constitute all there was of medieval society. In some areas, agriculture was organized on a nonmanorial basis. Moreover, the towns and the cities of the ancient Roman Empire often survived, though much diminished. From the eleventh century, they began to grow again. This growth occurred earliest in Italy and the Netherlands, and spread from there to other parts of Western Europe, producing numerous important urban centers. London and Paris, Lübeck and Naples, Bruges and Bergen, and numerous other cities, made their distinctive contributions to medieval life.[23]

This system began to shift from the eleventh century to the thirteenth, during the High Middle Ages, when population and the amount of land under cultivation rose and new tenure arrangements increased. Primarily, the number of freemen increased and obligations of service were transformed into monetary payments. Similarly, the lord's monopoly rights over proto-industrial enterprises such as milling, brewing, and making cloth were also monetized. Following the wars and plagues of the fourteenth century, depopulation led to a contraction of the amount of cultivated land and to further adjustments in the relations between lord and peasant. To some extent, an excess of land placed the peasant in a better position to live independently, and more freemen purchased or otherwise earned their own land.[24]

A great deal of this structural breakdown of the manorial system derived from advances in technology and an increase in other potential vocations, many of which were based on new or evolving energy exchanges. Mechanical, animal, wind, and water power had been applied to mills for doing laundry, tanning hides, sawing wood, casting iron, mashing pulp for paper making, and operating fullers' vats, bellows for blast furnaces, and hydraulic hammers in foundries. By the fourteenth century, the lathe, the brace and bit, and the spinning wheel had come into general use as methods for transferring motive power drawn from humans, animals, or water.

The era known as the Glorious Revolution was particularly important in the evolution of capitalistic, surplus-generating arrangements in English agriculture.

Expanding on basic windlass technology after 1800, more intense industrial purposes could add heavier chains and mechanized power sources. LIBRARY OF CONGRESS PRINTS AND PHOTOGRAPHS DIVISION

For instance, the Declaration of Rights signed after the revolution by William of Orange was a declaration of the rights of a Parliament dominated by progressive "bourgeois" landholders who were engaged in exploiting new techniques in speculative, market-oriented ventures. From this point forward, the British Parliament and courts grew increasingly dominated by capitalistic agricultural and commercial interests, and they guided a national economy toward advancing this end.

TECHNICAL INNOVATION EXPANDS AGRICULTURE

Particularly in Europe, without diverging too significantly from a reliance on the muscular exertion of people and animals or by recycling organic wastes and planting legumes, agricultural efforts increased by applying technologies such as the more efficient use of animate power or greater intensity of cropping practices. However, more and more farmers expanded their use of animate power capabilities as well as the integration of the kinetic energy of water and wind to steadily modernize agriculture in Europe and later in the Americas.

In terms of the agricultural revolution, the period 1500–1750 is often organized with the term *new husbandry*, and it appeared first in the Low Countries and slowly spread to England and eastward. Still, by 1750, adoption of these practices had barely begun in some countries, including France. The expansion of three basic practices can be traced during this period: stall-feeding of cattle, new crops, and the elimination of fallowing. The results of this basic shift included more and better fed cattle, which increased the supply of animal products and produced more fertilizer, which helped to increase crop yields. This allowed farmers to introduce new fodder crops, including alfalfa, clover, turnips, and other grasses, that improved soils, enhanced nitrogen levels, and broke disease and pest cycles. While the results were not exactly a "revolution," output, reliability, and standardization increased significantly.

The period that is often viewed as a truer agricultural revolution took place in England during the 1700s when more intensive use of land became possible through the substitution of crop rotation for fallowing. Farmers of this era began using "artificial" legume fodders that replaced the fertility of the soil drained by traditional cereal crops and provided additional fodder for cattle. This shift effectively broke the dung–fodder cycle, providing more fertilizer to improve soil fertility, which, in turn, produced even more fodder, cattle, and fertilizer. There were other improvements, but this was the foundation of the surplus in agriculture that fed the Industrial Revolution in late eighteenth-century England. The way had been prepared by the evolution of feudalistic into capitalistic agriculture, by an infusion of the commercial spirit, and by prior elaboration of market arrangements. Over centuries, property rights in land had evolved toward individual and absolute tenure.

After the manorial period, enclosure had been enforced against common use, and land had been divided in sufficiently large units to accommodate the demand for foodstuffs in industrial towns. In England, where defeudalization had gone furthest, the agricultural revolution first occurred. Thereafter, while England turned to manufacturing, similar agricultural advancement occurred in France. The growing success of agriculture produced a surplus that might be used for investment, which occurred as the institutions of feudalism were eliminated. This was achieved almost two centuries after settlement began in French and British American colonies. The French Revolution followed hard upon the American Revolution, and both entailed formal abolition of feudal tenure. The difference was that significantly greater informal shifts in these institutions occurred in England and New England than in New France. To be clear, the actual practices of agriculture had changed little. Certainly, there was a reduction in the time that men and women spent in the fields. Most of the new techniques saved money and land but increased labor of other sorts. The major shift, though, occurred in the reliable nature of food production that could be carried out by only a segment of the population, thereby leaving labor and investment available to pursue other endeavors.

Involving animal labor in agriculture, particularly plowing, flourished during this period of agriculture. The use of plows, yokes, and harnesses that would direct and capture the movement and power of oxen and cattle generally dates back to oldest Mesopotamia and was common very early in almost every region of agricultural advancement. Both the breast-band harness and the collar harness are considered to be Chinese inventions, where both cattle and horses were used in agriculture.[25] Because of the concentration of cattle's weight in the front, Smil writes that horses, "whose body mass is almost equally divided between the front and rear," were the most powerful draft animal. Regardless of the animal, the energy cycle for their agricultural use was similar: the tradeoff of their use was the need to also grow grain for feed. As horses became more of an emphasis for work as well as for transportation, many technological innovations were required to inculcate their use deeply into human culture. The primary reason, though, that cattle remained the most common animal for field work in China, Africa, and India was their ability to survive on roughage.[26] As ruminants, cattle ate primarily straw that could be gotten from informal grazing, making them, in Smil's words, a "clear energetic bargain."

HORSESHOES HELP TO HARNESS ANIMAL POWER

One of the preconditions of the horse's success as a work animal was protection of the animal's softer hoof. Humans protected the horse's hoof from excessive

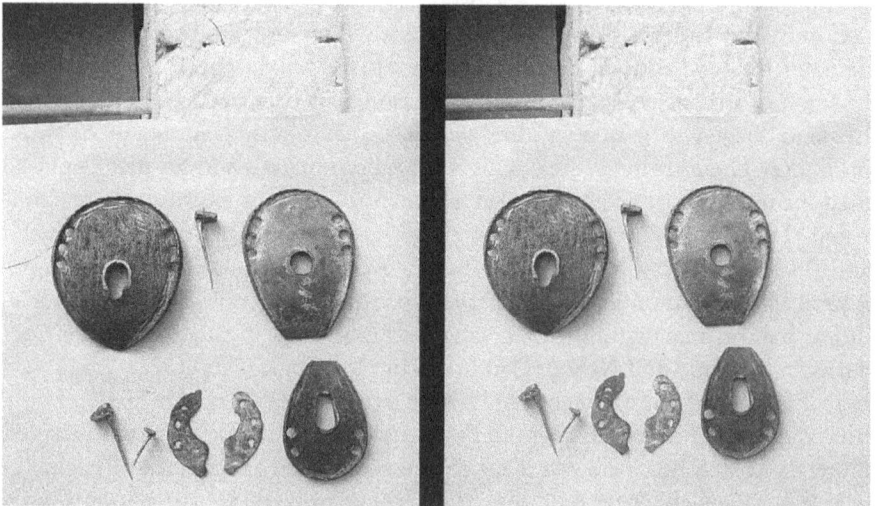

Animal power could be systematized and extended through innovations such as horseshoes. LIBRARY OF CONGRESS PRINTS AND PHOTOGRAPHS DIVISION

wear by attaching to it a narrow, U-shaped metal plate known as a horseshoe. Small nails held the shoe in place on the horses' insensitive hoof and allowed the animal better traction and endurance. There were some variants: Greeks attached leather sandals to each hoof, and Romans used clips instead of nails. However, particularly in the cool, wet climate of Western and northern Europe, horseshoes were critical to harnessing the animal's power.[27]

CURRENT: Bringing the Products of the Countryside to Market

In every aspect of the agricultural economy that took place in England, the most distinguishing advancements grew out of systems of exchange that were unique. Creating methods for trading linked the countryside to towns and cities and also made farmers able to support themselves more easily and create a surplus that could be traded or sold through one of the new outlets. The agricultural animals provided a vital component of this exchange when English agriculture shifted to the era of enclosure and herding became more of a business enterprise. By the 1400s, herders led tens of thousands of sheep to be sold at markets throughout the countryside. The sheep fair offered men who spent long periods of time alone with their animals to spend time with comrades. The fair became a time of merriment for shepherds and also a rare moment of respect when their flock was judged worthy of trade. In the end, though, the energy exchange still originated in pasture grasses, grown by the Sun, that were then fed to animals in an increasingly efficient and specialized system.

Animal and produce sales were essential to organizing English fairs. For instance, in fairs in the Wessex area of England, each day of the fair was assigned to a specific animal or crop: day four was sheep day. Other days were assigned to cattle, horses, and corn. At such fairs, as many as five hundred thousand animals might be traded during the length of the fair. By the 1500s, the system of livestock fairs knitted together a broad English countryside. Scottish drovers bound themselves to accessible English fairs, including those held in London. Many Scots felt that this undercut Scottish markets. After the crowns were joined in 1603, this practice was no longer frowned on. The Irish and Welsh drivers also shared the English markets by the 1600s.

In the early feudal period, goods were produced in very limited number and dispersed and sold over only a small area. Typically, any products needed to be used immediately and, therefore, could not be sent far away. By the late 1300s and 1400s, however, private production and sales had begun to liberate individuals from the limits of feudal life where wealth was concentrated with lords and royalty. Markets and fairs offered the start of such changes in Europe. Although many early markets involved barter and trade, they also ushered in the use of currency for exchange. Barter and trade systems, of course, significantly limited the ability of traders to disperse goods. The slow introduction of currency all over Europe was a necessary precursor for further trading. The use of money or gold increased the flexibility of trade, which made exchange between regions and, eventually, between cultures more possible.

During the 1300s and early 1400s, the peasant family might still produce its own food and clothing, but now they required money to pay rent or to buy farming tools. The lords and monarch, then, required money on a massive scale. New, long-distance trade meant exotic luxuries could be obtained at a price. The creation of a consumer culture also stimulated further use of currency and brought trade networks deeper into Europe by the 1400s. The emphasis on gold and currency also fueled the desires of European royalty to expand resources by military means. Possessing strong armies meant the ability to conquer other nations and the ability to condense wealth. Territorial expansion and the search for new supplies of gold fueled efforts to navigate the open ocean and to liberate Europeans from the limits of land travel.

These were the general transitions at work in European societies that largely kept machines and technology from developing fluidly from 1300 to 1650; instead, great intellectual changes were needed in European society to create a culture that was willing to pursue innovations. When this culture took shape in the 1500s and 1600s, a period of unprecedented technological expansion followed. By approximately 1200 the economies of Western Europe had absorbed many of the technological and cultural innovations that Islam and the East had to offer. Ideas, techniques, and even actual inventions were brought to Europe by way of trade networks. By building on ideas drawn from great societies of the world, European society surged ahead of every other region during the three centuries that followed.

New ways to develop ideas and technology into profitable opportunities defined the era after feudalism and led to the rise of guilds and craftsmanship, which also stimulated invention. With these economic developments behind them, cities grew rapidly in medieval Europe and fueled new economic growth and the revival of trade. Most of these towns and cities owed their urban status to the presence of military garrisons during the Middle Ages. They were not very large and were distinguished from the surrounding countryside largely by the possession of protective walls. In the medieval era, however, trade networks enabled towns to grow at a rate unknown in previous centuries.

Many new settlers made homes beyond the town walls, which forced townspeople to extend existing defenses or to alter their approach to the outside world. Instead of walls to separate them from the rest of the world, economic opportunities and new ideas began to link people together. Not only did the size of the towns increase but a whole new way of life came to be established within them. Many of the inhabitants were merchants who used the city to conduct their trade, often international in scope. The expansion of trade ushered in new economic possibilities for those who employed technology.

For instance, Europeans by the 1500s had encountered ideas of the Mayans of Central America, who were advanced mathematicians and astronomers. The Mayans developed the most complex writing system of Mesoamerica, which used pictographic symbols (images whose meaning is communicated through their visual form). The Incas of South America had precise mathematical notations, were accomplished astronomers and builders, and practiced sophisticated medicine, including brain surgery. The Incas were also advanced metallurgists who had lined the Temples of the Sun and Moon at Cuzco in Mexico with gold and silver. Through exploration, trade, and colonization, Europeans transferred the knowledge of other cultures to their own shores. In Europe this knowledge was utilized in the evolving economic system of industrialization that was being initiated in pockets with available capital and investment.

PRIME MOVERS STIR PROTO-INDUSTRY

Metallurgy, particularly based in the use of furnaces and bellows, is a proto-industry normally dated to Song China of the eleventh century. These industrial sites marked one of the first examples of the complex joining of different prime movers into the creation of a product. Built into hillsides, the early Chinese blast furnaces began with charcoal and eventually burned coal and coke. The enterprises also used water power for various aspects of industry.[28] As this technology transferred to Europe, changes in manufacturing from 1300 to 1650 brought major alterations to the economic organizations of European society as well as the availability of goods and services. The wars of the Renaissance and Reformation era proved to be a great boon for merchants and manufacturers supplying armed forces. Many of these new industries and systems of transportation would ultimately be put to peacetime uses as well. However, by most modern measures, the manufacturing that took place from the 1300s to the 1500s was of a very limited scale. Between 1500 and 1750, changes in manufacturing continued but would not accelerate remarkably until after 1750. Historians often use the term *proto-industry* to describe most enterprises of this era because, by and large, they occurred as freestanding experiments. A common start for each undertaking, though, was energy.

In the era of proto-industry, humans were the base-level prime mover for any enterprise. Beyond that, part of the origin of any enterprise involved not only how to perform a desired task but also how to harvest a source of power for the purpose that might be sufficiently reliable. By and large, the manufacturing that did develop was most often based on technologies that European merchants brought from other regions, particularly Asia. For instance, Europeans perfected the art of making porcelain imitations of Chinese crafts. And from India, Europeans imported methods for manufacturing silk and textiles. While perfecting these technologies, European business leaders also linked specialized craft production into larger-scale systems that also placed small-batch into the class of manufacturing.

With each successful enterprise, lessons were learned and shared. Through this cultural exchange advancements were devised that formed the basis for this system of manufacturing and advanced the use of energy resources. Ultimately, the outcome was a large-scale shift in economic and social patterns in Europe that culminated with the formation of an entirely new organization to society.

At its most simple level, technology can mean innovative practices and ideas. More than in any other era in human history, the Renaissance period saw innovation take material form as devices that would become known as "machines." For nearly every device, a source of power was required—a prime mover. Early technologies relied on the first prime mover: muscle power. Windlasses, capstans, treadwheels, and gearwheels were critical to lifting, grinding, crushing, and pounding. Treadmills were designed to exploit the largest muscles in the human, which are in the back and legs. Typically, a treadmill or other generator transmitted power to gearwheels and could be used for a variety of tasks, including pumping water. Internal vertical and horizontal treadwheels could also be adapted for use by animals, usually horses, which could produce more power than humans. In the seventeenth century, strong animals were commonly used for grain milling and oil pressing. Later, horses were used in various industries that required rotary power, including raising coal, ore, or water in mines. By the 1700s, however, most mills had shifted to more easily manageable sources of power, such as wind and water.

Wind Power

Combining the ideas of the watermill and the sail, the windmill is generally attributed to Muslims from central Asia. The first windmills were noted in Europe before 1200. The Domesday Book of 1086 counted 5,624 mills in southern and eastern England—an estimated one mill for approximately every 350 people. The early mills were generally fitted with a pivoting windmill that could be moved to accommodate Europe's variable winds. These "post mills" were unstable in heavy winds and storms and their peak performance was limited by their small size.

In its general design, the windmill's infrastructure came from its sister technology, the water mill. The horizontal axle and mechanisms for gearing and transmission were identical to those used in water mills. This contrasted with most mills in Asia, however, which did not use gears or a horizontal axle. Windmills in Europe by approximately 1500 were normally tower and smock mills, which were both designed so that only the top cap was turned into the wind. Smock mills were normally made of wood, and tower mills from stone. In 1745 English operators introduced a device known as a "fantail" to power a winding gear to turn the sails automatically into the wind.

Dutch millers were the leaders in wind power although they did not add fantails until the nineteenth century. Dutch millers added a canted edge to the blades or sails, in contrast to the flat edge used by other millers. The Dutch blades gained more lift and reduced drag. Cast metal gearing and governors, which were used to limit the blades' speed on windy days, were also added by the Dutch millers to prevent the canvas from spinning too fast. At the start of the medieval era, post windmills and water mills had a similar capacity to generate power, yet by the end of the eighteenth century most waterwheels were four to five times more powerful than the best windmill.

Water Power

Vertical wheels were first used to capture the motive power of rushing water. Most often, the force was used to power a millstone to grind wheat and other grains into food material, including flour. These machines could be used only in a limited number of locations; early improvements in the technology by approximately 1400 attempted to make waterwheels more adaptable to variations in location and water flow (for instance, using man-made waterways to deliver water to mills away from rivers).

Waterwheel use gradually expanded beyond grain milling to replace a number of other manual tasks including oil pressing, sawing, wood turning, paper making, tanning, ore crushing, iron making, wire pulling, stamping, cutting, metal grinding, polishing, and blacksmithing. In each case, work was usually carried out on a fairly small scale and water power was substituted for human or animal labor, which could not sustain the continuous and reliable power of water. Waterwheels also opened up other productive possibilities and became the mainstay of early industrialization in Europe.

Tidal Power

Tidal mills were a medieval invention that was first mentioned in the twelfth century in both England and France. These mills were built in low-lying areas near the sea. Additionally, dams with swinging gates were built along shallow creeks so that as the tide came in, the gates swung open inward, away from the sea, and water filled the area behind the dam. When the tide turned, the gates swung shut, forcing the water to flow seaward through the millrace of the tidal mill.

The obvious disadvantage to tidal mills is that the time of the tides shifts every day. Thus the millers had no choice but to work hours dictated by the tides. These mills seem only to have been used to grind grain. There were never very many of them compared to the number of waterwheels and windmills.

FORMALIZING PROTO-INDUSTRIAL ENTERPRISES IN EUROPE

From scattered, proto (or confined) beginnings, a few industrial technologies lent themselves to systemization and scaling that allowed them to graduate to enterprises that could support entire communities—not only to supplement agricultural towns. This proved the important step toward the "takeoff" that could use energy to grow into full-blown industrialization.

Water-Powered Milling

Of these early prime movers, waterwheels proved to be the most expansive application for manufacturing in the early industrial era. Typically, the primary device for capturing the river's motive power was a large wheel, known as a breast wheel. Water and gravity functioned to spin the wheels and a close-fitting breast work steered the water into the wheels, which were most often designed so that the water entered below the level of the center shaft. These wheels were referred to as undershot wheels. Overshot wheels, in which the water hit above the shaft, would also come into eventual use.

Undershot wheels could be placed directly in a stream, which made them simpler to site but also more prone to flooding. Overshots needed a regulated water supply and, therefore, involved additional construction of flumes, ponds, and ducts.

Whichever type of wheel was used, the motive power was then transferred into a mill through the use of gears. Through the eighteenth century, wheels, gears, and shafts were most often made of wood. Afterward, the more durable iron would take over as the primary construction material for manufacturing hardware.

Throughout the 1600s, wheels remained limited to producing approximately 4 kilowatts. The only way to expand a manufacturing site's capabilities was to add wheels. For instance, by the late 1600s, one of the best-known industrial sites was the series of fourteen large waterwheels on the Seine at Marly that were built to pump water for the Versailles fountains of Louis XIV. Each wheel could muster only 4–7 kilowatts.

Burning Wood and Considering Coal

During this early period of industry, Western Europe's forests had been severely diminished as the raw material used in shipbuilding and metallurgy. Foreign wood allowed shipbuilding to continue. English ironmasters, however, discovered a new source of energy that would greatly multiply the scale and scope of industrial potential. The English use of coal and more specifically coke revolutionized manufacturing.

Although Western Europe had abundant supplies of ordinary coal, it had proven useless for smelting ore. Its chemical impurities, such as phosphorus, prohibited its generation of strong iron. For this reason, smelting was fired with charcoal, which was made from wood.

Western Europe's lack of wood made it lag behind other regions during these decades. However, in approximately 1709, Abraham Darby discovered that he could purify coal by partly burning it. The resulting coke could then be used a smelting fuel for making iron. Darby released this knowledge for public use in 1750.

Power to Make Iron

A single product can demonstrate the changes in manufacturing better than any other commodity. Iron had been known since antiquity in many parts of the world, but particularly in Asia. Although iron production became very important, the difficulty in producing pure iron remained until late in the Middle Ages. As a fledgling manufacturing enterprise, iron forges joined a variety of the small-scale prime movers to form one of the earliest industries in human history.

Waterwheels powered the bellows used in various English monasteries in the twelfth and thirteenth centuries and later spread over much of Europe. The mills also powered trip-hammers to pound the spongy bloom to wrought iron. By 1350 blast furnaces had been invented, and they spread throughout Europe during the 1400s. Typically, these furnaces were fired with wood or charcoal. This more intense heat resulted in a very hard and strong but brittle form of iron called *cast iron*. But since it was now possible to produce the high temperatures needed to actually melt iron, cast iron could be transformed into steel, hundreds of pounds at a time. By 1500, Europe was producing upwards of sixty thousand tons of iron per year. All of this production pales in comparison to the ability of iron producers once they shifted to coal or coke to fire their furnaces; however, it marked an important start and a transition from proto-industries to enterprises of a larger scale and scope.

CONCLUSION: THE SETTING FOR EXPANSION

The Middle Ages in Europe were marked by a reluctance to develop commerce beyond the local markets. When European nations reached outward to convert others and spread their religion and beliefs, these "crusades" also represented a new contact with other societies that eventually led to trade networks with Arabic cultures and others. Beginning around 1200, this trade also influenced new European economic thought that would eventually result in the concepts known as capitalism and mercantilism. First, though, the new economic views required a different worldview: Europeans looked outward in a fashion not seen before toward the riches of other regions that had prospered. The same curiosity that led researchers to burst through scientific frontiers during the Renaissance and Reformation also sent explorers and traders all over the globe.

The first breakthrough for Europe was accessing the southern trade networks that had taken shape in a fairly limited region around the Indian and Mediterranean Oceans. During the 1200s, European explorer and writer Marco Polo traveled into the region and wrote of the Silk Road that tied together the Arabic and Islamic world in a system of cooperative trade. His description of the rich civilizations taking shape along this early trade network stirred new European interest in trade and economic development. Europe's first connections with what is referred to as the "southern" trading world emanated from Italian cities, which pooled wealth to establish some of the first banks. These Italian cities also established small industries to take advantage of the raw materials (such as silk) available from the East. Similarly, Portuguese traders led a European rush into other areas of Africa, India, and Asia. Beginning in the 1400s, new technologies—particularly military abilities such as a portable cannon and maritime know-how and shipbuilding—allowed European powers to enforce economic monopolies on products in distant lands. This success led to a period of oceanic exploration in the 1400s that was directly tied to trade expansion.

The first connections between regions for trade also brought the difficulties of mixing biological materials. By the beginning of the fourteenth century, the population had grown to such an extent that the land could provide enough resources to support it only under the best of conditions. There was no longer any margin for crop failures or even harvest shortfalls. At the same time, however, the Western European climate was undergoing a slight change, with cooler and wetter summers and earlier autumn storms. Conditions were no longer optimum for agriculture. Into this precarious setting, new trade patterns brought a disease-producing bacterium, the bubonic plague. Marks deems that because of the various factors of humans during the 1300s, "the plague was not a purely 'natural' phenomenon."[29] Moving through the shipping stores of the period, infected fleas on mice are credited with the spread of what became known as the Black Death, which historians believe reduced the global human population from 480 million to approximately 350 million.

The world in 1400 had begun to diversify; however, the majority of Earth's 380 million humans continued to live in the countryside. The world's twenty-five largest cities in 1400 (ranging in population from 500,000 to 80,000) were only responsible for approximately 1 percent of the global population. Nine of these population centers were in China, including the world's largest city, Nanjing. In ranked order, the next largest cities were Vijayanagar (south India) and Cairo before finally reaching a European city at fourth, Paris.[30] Although the world's wealth was concentrated in Asia, clear shifts were underway.

Along with the growth of urban areas came changes in banking and in the technology that supported manufacturing. A class of big businessmen arose and in connection with it an urban working class, or proletariat. For this new urban society, new types of legal institutions and property tenure had to be devised. A mercantile law, or law merchant, grew up to settle cases arising from trade disputes. Property holding was set free from the complex network of relationships and obligations that had burdened it, and it became possible for city dwellers to hold property outright. One of the most distinctive characteristics of urban life was new freedom that had not been seen in the feudal countryside. Towns grew and flourished; trade, banking, and manufacturing became established on a new scale; more and more persons achieved the legal status of free men. To accommodate these changes, vast tracts of land, which had been uninhabitable forest or swamp, were cleaned, drained, and subjected to cultivation. A new order and urgency came to the landscape of production and industry.

In the unique city of Amsterdam, specializing in dam and canal building in order to manage its own predominance of swamplands, people of the Low Dutch Countries lived in a system that was not restricted by the feudal system but on individuals who bought and sold property and formed mutual partnerships. It was partly this spirit that led Amsterdammers in 1599 to meet atop one of these dams or dikes to form the world's first multinational corporation, the Vereenigde Oostindische Compagnie, the United East India Company, which soon became universally known by the logo of its interlocked initials that would appear on ships in harbors around the world: VOC.[31] Along the canal known as Damrak, business leaders invited community members to buy stock in VOC through 1602 and, then, those stocks that would become some of the most valuable in human history could be traded. Shorto writes:

> Like the oceans it mastered, the VOC had a scope that is hard to fathom. One could craft a defensible argument that no company in history has had such an impact on the world. . . . In innumerable ways the VOC both expanded the world and brought its far-flung regions together. . . . It pioneered globalization and invented what might be the first modern bureaucracy. It advanced cartography and shipbuilding. It fostered disease, slavery and exploitation on a scale never before imagined. It shuffled the global ecosystem—by design and by accident. . . . The genius of the VOC was in threading itself through this highly evolved network. By the end of the golden age a century later, the Dutch were selling spices not only to Europe but to China, India, and even to the Spice Islanders.[32]

In short, Amsterdammers had realized the next revolution would come from the connections between nations and regions. Although productive capabilities themselves were not yet changing dramatically from the biological old regime, new worldviews and the connections enabled by wind power altered the world dramatically after 1600.

NOTES

1. Russell Shorto, *Amsterdam* (New York: Vintage, 2013), p. 29.
2. Shorto, p. 45.
3. Shorto, p. 102.
4. Jared Diamond, *Guns, Germs, and Steel* (New York: Turtleback, 2005), p. 93.
5. Alfred Crosby, *Children of the Sun* (New York: Norton, 2006), p. 5.
6. Crosby, p. 2.
7. See, for instance, David R. Montgomery and Daniel J. Sherman, *Environmental Science and Sustainability* (New York: Norton, 2021), pp. 70–72.
8. Diamond, pp. 103–105.
9. Diamond, pp. 103–105.
10. Diamond, p. 450.
11. Diamond, pp. 103–105.
12. Crosby, pp. 46–49.
13. Vaclav Smil, *Energy and Civilization: A History* (Cambridge: MIT Press, 2017), p. 33.
14. Crosby, p. 12.
15. Crosby, p. 12.
16. Crosby, p. 13.
17. Crosby, p. 32.

18. E. A. Wrigley, "Energy and the English Industrial Revolution," *Philosophical Transactions of the Royal Society A.* 371, no. 1986 (2013). In this article, Wrigley argues: "The classical economists identified three basic requirements for production to take place: it must involve capital, labour and land." Readers may also wish to consult E. A. Wrigley, *Energy and the English Industrial Revolution* (London: Cambridge, 2010), and *The Path to Sustained Growth* (London: Cambridge, 2016).

19. Smil, p. 132.

20. See, for instance, Joel Mokyr, *The Culture of Growth* (Princeton, NJ: Princeton University Press, 2018).

21. Smil, p. 134.

22. Smil, pp. 138–139.

23. See, for instance, Bill Gilbert, unpublished manuscript "Renaissance and Reformation," available at http://vlib.iue.it/carrie/texts/carrie_books/gilbert/.

24. Astrid Kander, Paolo Malanima, and Paul Warde, *Power to the People: Energy in Europe over the Last Five Centuries* (Princeton, NJ: Princeton University Press, 2014), pp. 86–90.

25. Kander et al., p. 70.

26. Kander et al., p. 68.

27. Smil, p. 73.

28. See, for instance, Donald B. Wagner, "Blast Furnaces of Song-Yuan China," *East Asian Science, Technology, and Medicine* 18 (2001), available at https://core.ac.uk/download/pdf/228877749.pdf.

29. Robert Marks, *The Origins of the Modern World* (Lanham, MD: Rowman & Littlefield 2019), pp. 37–39.

30. Marks, p. 24.

31. Shorto, pp. 102–103.

32. Shorto, pp. 103–104.

TRANSITIONING BY THE NUMBERS
Biological Old Regime

As humans moved from the biological old regime, their energy use continued previous patterns of exchange while enhancing their capabilities with limited use of new methods and options.

Energy Capacity by Source

Ranking	Examples	Energy Density (MJ/kg)
Foodstuffs		
Very Low	Vegetables, fruits	0.8–2.5
Low	Tubers, milk	2.5–5.0
Medium	Meats	5.0–12.0
High	Cereal and legume grains	12.0–15.0
Very High	Oils, animal fats	25.0–35.0
Fuels		
Very Low	Peats, green wood, grasses	5.0–10.0
Low	Crop residues, air-dried wood	12.0–15.0
Medium	Dry wood Bituminous coals	17.0–21.0 18.0–25.0
High	Charcoal, anthracite	28.0–32.0
Very High	Crude oils	40.0–44.0

SOURCE: Vaclav Smil, *Energy and Civilization: A History* (Cambridge: MIT Press, 2017), p. 12.

Estimates of Energy Capacity in Each Animal Source of Power

Animals	Common Range	Large Size	Typical Draft (kg)	Usual Speed (m/s)	Power (W)
Horses	350–700	800–1000	50–80	0.9–1.1	500–800
Mules	350–500	500–600	50–60	0.9–1.10	500–600
Oxen	350–700	800–950	40–70	0.6–0.8	250–550
Cows	200–400	500–600	20–40	0.6–0.7	100–300
Buffaloes	300–600	600–700	30–60	0.8–0.9	250–550
Donkeys	200–300	300–350	15–30	0.6–0.7	100–200

SOURCE: Vaclav Smil, *Energy and Civilization: A History* (Cambridge: MIT Press, 2017), p. 67.

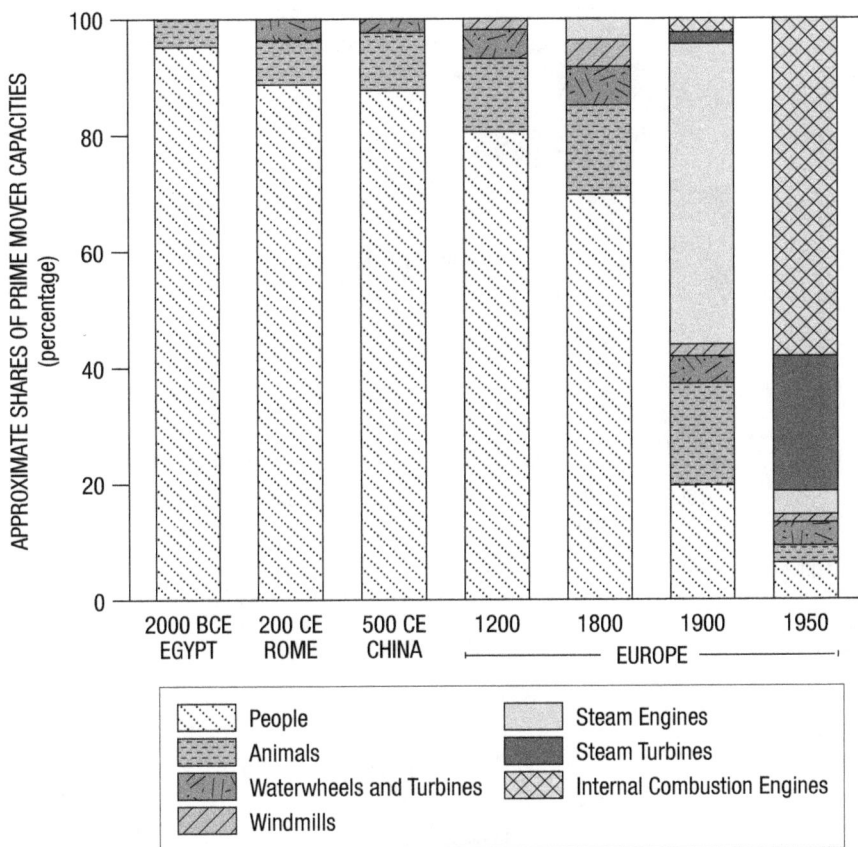

Approximate Shares of Prime Mover Capacities. VACLAV SMIL, *ENERGY AND CIVILIZATION: A HISTORY* (CAMBRIDGE: MIT PRESS, 2017).

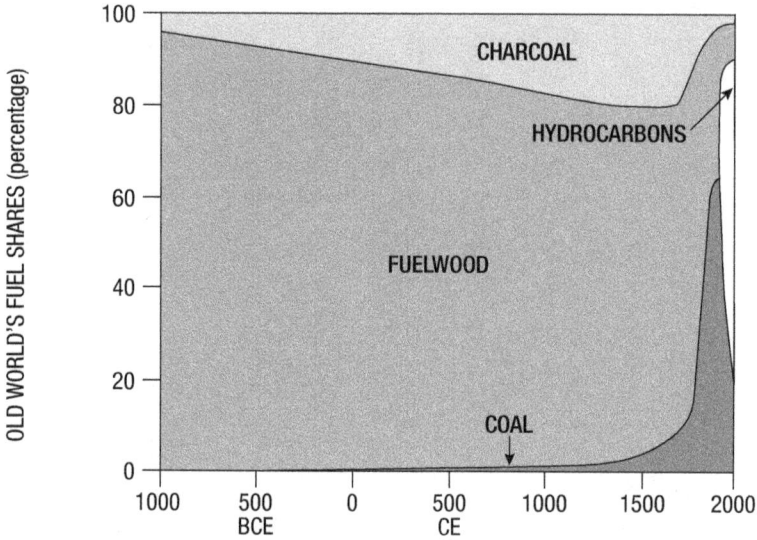

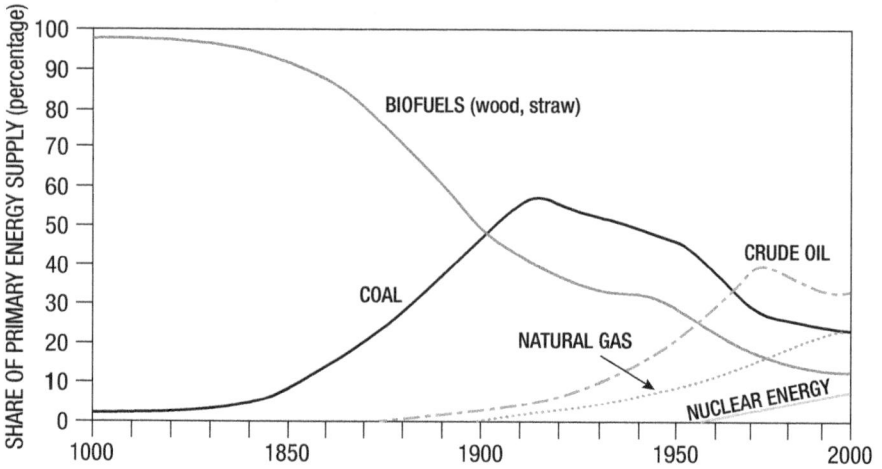

Top: Old World's Fuel Shares. *Bottom:* Share of Primary Energy Supplies. VACLAV SMIL, *ENERGY AND CIVILIZATION: A HISTORY* (CAMBRIDGE: MIT PRESS, 2017).

II

INDUSTRIALIZATION AND THE GREAT REVERSAL (1400–1920)

There was no particular eureka moment to ensure that humans learned to burn fossil fuels in order to release stored energy. Instead there was a steady stream of innovations teamed with a consistently growing willingness to derive work and labor from stored reservoirs of energy instead of from actual bodies, such as animals and humans. Overall, the greatest single change brought by this shift, which is normally referred to as the Industrial Revolution, was the release of human accomplishment from the limits of the caps on human and animal energy by techniques and devices such as this blast furnace of the early 1800s.

In this image, a worker "charges" the furnace with soft coal in Tipton of the West Midlands of England. From its neighboring Black Country towns, the raw materials, iron ore and limestone, could be gathered to run an early iron manufacturer to make hinges, screws, awl blades, tools, and nails. In particular, Tipton's population made some of the world's best-known nails, which were used for all kinds of construction.[1]

NOTES

1. http://www.historywebsite.co.uk/articles/Tipton/ironworks.htm. Accessed on January 6, 2022.

2

Colonialism, Mercantilism, and Empire

GROUND: LEARNING THE ROPES

At the heart of maritime society was the sailing ship, the great technological marvel of the early 1800s. As an energy technology, though, sailing required a significant supporting infrastructure of know-how and tools. For commercial ships, this system was focused in ports and shipyards. Typically, the shipyard was owned and operated by a master shipwright. In the United States, this businessman met with the prospective customer (typically an investor who viewed the ship as a potentially profit-making mechanism) to establish what type of vessel was desired: sloop or schooner for the coastal and West Indies trades; brig, brigantine, or a full-rigged ship for Atlantic crossings. The builder and customer would then settle on a size and a price per ton.

The shipyard was stacked with the proper wood for each portion of the sailing vessel. White oak was most popular for the keel, stem, sternpost, and frames. Hackmatack was for the knees that supported each level of the ship, and oak or pine for the planks and decking. Specialized construction techniques were then required to transform the wood into an operational craft. Based on techniques perfected all over the world since the 1300s, craftsmen, including joiners, caulkers, and smiths, then built the vessel. Finally, in order to protect the hull, it was covered in a mixture of pitch and lime. To become operational, the wooden tub now needed to be enabled to harness its power source: the wind. To connect to its power source, the finished hull was then rigged out by riggers, sailmakers, and block makers before the shipwright turned the vessel over to its owner.

The organization of the vessel's sails indicated the ship's type, its use, and even how far it might be capable of traveling. Intercoastal ships carried a great deal of materials from one American port to another. Typically, these vessels, which included sloops and schooners, flew a number of sails but emphasized one central, triangular sail. The

43

triangular sails derived from Asian models and were considered to be more flexible
for finding changing winds. However, ships and barques traveled across vast stretches
of ocean. The equivalent of the modern-day tractor trailer, these vessels were rigged
with a system of square sails.

The size of vessels increased throughout the 1800s. By 1850, shipbuilders had
proven the cost-effectiveness of the triple-masted clipper ships that carried more sail
than any other vessel. Of course, this also required more crew to operate. As the Age of
Sail reached its apex after 1850, it succeeded because of a system of trade and opera-
tion that began with wind power but relied on the clipper more than any other vessel.

The order from the merchant in Liverpool left little doubt about the purpose of
the product being requested: Joseph Manesty wanted two ships "for the Affrica
trade." On August 2, 1745, he wrote to John Bannister of Newport, Rhode
Island: "No trade [was] push'd with so much spirit as the Affrican and with great
Reason—high profits!" But, he continued, "ships are so scarce here that none is
to be had at any rate."[1]

The perfection of the trade system over the Atlantic Ocean reached its aw-
ful pinnacle with the slave ship, which through the lens of energy exchanges
becomes a predecessor of the modern-day oil tanker. Although slavery had been
previously used throughout the world, the bodies carried by these ships during
the 1700s powered the plantation agriculture that allowed mercantilism to overcome limits of climate and markets.

The exploration of Christopher Columbus has been one of the basic tropes of
global history since educators began teaching courses on "Western Civilization."
Overcoming the positive bias that defined a few centuries of Columbus instruc-
tion, however, is only part of the difficulty with this point of origin. While we
have come to see European biological and economic expansion—personified
by Columbus—as a conquering act that unduly influenced indigenous history
throughout the globe, we haven't yet entirely performed the "systems-level"
analysis that fully places Columbus as a minor actor in a massive shift in the
human species.

In a systems-level, macro analysis, the best symbol of this era of energy is not
a single, bold explorer such as Columbus but, instead, the slave ship. Within
the morally barren hull of the sailing ship, one sees the emergence of an energy
system that grew from primitive technologies: harnessing wind power and organizing labor of the human body. Each ship functioned as a portable energy
source to be applied as desired. The logic and organization, then, became the
crucial element in exploiting how this energy might be applied to strategies of
national development.

Agriculture had clearly enabled humans all over the world to apply the Sun's
energy to a routine existence of various sorts by 1400. In these distinct pockets

of civilizations organized around exploiting Sun energy through agriculture, sharing of ideas and crops occurred slowly. It was an extension of the energy of the Sun—not the power of the human or animal body—that brought societies together. Crosby continues:

> The Sun wherever it shines heats bodies of air differentially. These bodies expend that sun energy scurrying and blustering about as winds, swerving in obedience to the Earth's rotation. Wind is a form of sun energy available on land and sea.[2]

While each sailing ship served as a critical opportunity for the expansion of human contact, the slave ship—holding its raw material of human power that would exploit the Sun wherever the ship landed—became the critical link in the first global economic system of trade made possible by the power of the wind.

Gutenberg, da Vinci, and other inventors lived in an era in which their efforts to create new technology were not always supported; at times, the search for new knowledge was even referred to as heresy. Galileo Galilei (1564–1642) discovered a number of natural laws (for instance, uniform accelerated motion, gravity, and oscillation) after a series of experiments with pendulums, inclined planes, and projectiles. Galileo, who is often credited with inventing the microscope, constructed a telescope, and observed the planets during this period, only to be condemned later for his views. Meanwhile, Johannes Kepler engaged in the study of astronomy and developed the laws of planetary motion in 1609. Each of these revolutionary thinkers, however, faced difficulty convincing others of his finding. In general, humans were just beginning to learn the worth of scientific exploration. An important part of the innovation of this era was the intellectual ability to question the boundaries of understanding.

Intellectual shifts initiated economic changes, and European society became increasingly willing to consider new ideas. Ultimately, economic development and the revolution in ideas of the Renaissance contributed to an increasing desire for new technologies. It was this revolution in thought that was essential to usher in the Industrial Revolution that began in the eighteenth century.

OCEANIC EXPANSION AND CULTURAL DIFFERENCE

The first global energy transition required obvious investment and emphasis from nations who wished to use it for a new network of interactions; however, as the system developed in the 1400s and 1500s, the energy being put to work was not an obvious, organizing mechanism. The sailing ship, as a technology, remained quite static until it became outmoded at the end of the 1800s. While there were variations in sail and ship design, the actual energy transaction remained, simply, to capture wind energy and direct it to satisfy humans' needs and desires. And, its severe limitations also remained the same.

In an era of outward expansion, national politics and culture played a significant role in how quickly each nation sought out new opportunities to apply to the sailing ship. Although Portuguese and Spanish sailors are credited with being the first explorers of this era, China had expanded its worldview well in advance of Europe (discussed briefly in the introduction). In the early 1400s, the Chinese emperor Zhu Di sent out the greatest navy the world had ever seen. Over fifty years before the first Portuguese explorers, fleets of hundreds of Chinese junks had already explored India, Arabia, and East Africa. Seven epic Chinese naval expeditions from 1405 to 1433 explored and brought under the Chinese tributary system the vast periphery of the Indian Ocean.[3] To lead these expeditions, Emperor Zhu Di chose Admiral Zheng He. In 1403, the emperor issued orders to begin construction of an imperial fleet of warships and support ships to visit ports in the China seas and the Indian Ocean. Bearing vast amounts of gold and other treasures, and with a force of more than thirty-seven thousand officers and men under their command, the Chinese explorers built great ships and set sail from Suchow.[4]

By midcentury, though, the Chinese view of this effort had changed wholesale. With increasing internal economic challenges related to extensive projects such as the Grand Canal and the Great Wall and a cultural belief that the nation's actions may have caused them "to be abandoned by heaven," when Zhu Di's son, Zhu Gaozhi, ascended to the throne upon his father's death in 1424, he immediately issued the following edict: "All voyages of the treasure ships are to be stopped." Not only were the ships ordered home, but the infrastructure of shipyards was ordered closed. As a result, by 1474 the fleet was down to one third of its earlier size. By 1503 the navy was down to one tenth of this size. The antimaritime party grew more powerful in China and made its power known through imperial edicts. In 1500 it was made a capital offense for a Chinese to go to sea in a ship with more than two masts without special permission. China's experience is the most dramatic example of the power of national authority over new efforts in trading.[5]

At this early moment in the Age of Sail, the motives of the Western explorers and the Eastern fleets were very different. The Chinese were essentially on a tour of the civilized world. They wished to conquer no one. Eventually, they wished to collect rich gifts of tribute from other nations. But their efforts were more symbolic than economic. European explorers, however, were engaged in a bitter war with Islam and an eternal search for gold and profit. Ultimately, China came to see overseas activities as an unnecessary economic drain. For them, economic considerations were reserved for inland activities. The European exploration, by contrast, expanded as nations competed against each other; the Chinese, however, acknowledged no competitors. Historian Marks credits these different cultural approaches to the outside world for China's abandonment of the sea in the 1400s. He argues that the politics of China simply turned inward, after years of debate and infighting to concentrate on internal development within its vast nation.[6] Europe, however, followed an entirely different model of economic development that built on the technology of harvesting and managing the wind.

CONDUIT: The Compass

For ships to attempt the large-scale navigation of the globe, instruments for navigation were essential. Originally a Chinese invention, the compass was first used for navigation after Europeans improved the device. Magnetized needles used as direction pointers are attested to in the eighth century in China, and they seem to have become common between 850 and 1050 as navigational devices on ships.

The first mention of the directional compass (as opposed to magnets themselves) occurs in Alexander Neckam's *De naturis rerum* (*On the Natures of Things*), which was probably written in Paris in 1190. Neckam's book was widely read by the end of the century, and by about 1218 Jacques de Vitry considered the compass a necessity for maritime navigation. By 1225 it was noted as being in use in Iceland; and in the Muslim world, the first mention of a compass is in a Persian story of 1232–1233. Typically, the early needles were made of iron and magnetized with a lodestone.

Nearly four hundred years later records note the first use of a compass to indicate direction by sailors for navigation. By the seventeenth century the needle was changed to become a parallelogram shape that could be mounted more easily upon the pin. In 1745 an Englishman, Gowin Knight, developed the Knight compass, which was an improvement because the steel needle retained its magnetism for a longer period of time than iron.

TECHNOLOGICAL DISTINCTIONS BETWEEN SEAFARING NATIONS

Although there were significant cultural differences between seafaring peoples, the technology that made sea travel possible was remarkably consistent. In fact, maritime travel was very much a joint technological evolution. At the time of the

Zheng He voyages, Chinese oceangoing technology was superior to European in most aspects—with the possible exception of navigation. Chinese ships were much larger—the largest ships of the Zheng He expeditions were about five hundred feet long. In contrast, the oceangoing European ships were typically a hundred feet long (with a weight of fifteen hundred tons for Zheng He's vessels and approximately three hundred tons for the Portuguese ships).[7]

The Chinese had been using multimasted ships for several centuries. In Europe, only Portugal had developed this innovation with its secretly designed "caravel." To power the devices, each society used the wind. The sails to catch and manage that wind, however, varied somewhat. Europeans sufficed with square sail rigs on their ocean vessels. These rigs were good for running across open distances; however, square rigs did not have the flexibility for moving against the wind and steering in close quarters. The Chinese had been using fore-and-aft lugsails (more efficient in beating upwind) since the ninth century. By the fifteenth century, however, the western and eastern sail technology was comparable.

The mariner's compass, so crucial to navigation in the open sea, was developed from the Chinese magnetized needle that had been developed centuries before. Compass technology traveled along land routes to the Mediterranean and was adopted by Europeans around the twelfth century. Therefore, both East and West had the mariner's compass in the fifteenth century.

Portuguese and Dutch explorers quickly gave Europeans an upper hand in their knowledge of wind and sea currents. Europeans also had a stronger knowledge of celestial navigation, which they had brought from Arabic cultures. The Arab and the Portuguese cross-staff or *balestilha* developed in the fourteenth century, and the astrolabe, by the early fifteenth century. The astrolabe provided better measurement of the angle of celestial objects.

By the sixteenth century, European mariners were clearly superior to those anywhere else in the world. This strength would help the seas to become a primary tool for European economic development.

Navigation Expands Trade Networks

Although navigation was still a relatively imprecise science, sailors were able to go farther and with more regularity than at any other time in human history. As the economy of Renaissance Europe developed, so too did the demand for imported goods and for new places to which to export local products.

Originally, the sailors of the Renaissance era took to the sea in order to supply Europeans with the many Asian spices, precious gems, and fine silk that were being purchased from foreign traders. The sea offered new networks of trade and new economic opportunities without the involvement of foreign traders. Essentially, European sailors could go directly to the source of the desired products.

Additionally, though, some mariners were drawn to the sea out of a curiosity to discover more about the world. These traders typically traveled on vessels carrying some of the world's first portable cannons. This allowed them to initiate a phase of "armed trade" in which coercion was freely used to open or enhance trade networks. These explorations led to trade for gold and ivory and, eventually, for slaves. Later, Portuguese sailors discovered the route around the southern tip of Africa that would take them to India entirely by sea.

Expanding Technology through Ship Design

Masts on a sailing ship function as pivots: the sail swings on them in order to catch the breeze. There were variations in the shape of the sail (normally square for long journeys), the number of masts, and the shape of the vessel's hull. European engineers adopted the lateen sail design (triangular in shape) from the Red Sea trading area and placed it on a three-masted ship known as a "carrack." The fully rigged ship combined the strengths of the lateen and square sails, which included maneuverability when sailing against the wind and extra speed when sailing with it. Overall ship and hull design also improved. The treadle looms (machines that wove thread into textiles) produced strong sailcloth. From China, European designers brought a sternpost rudder design to replace the clumsy steering oars used previously. Finally, Europeans adopted the "carvel" construction technique, which placed planks edge to edge with caulking (sealing material) between them to preserve water tightness. This technique saved on wood and made ships lighter and larger than previous types.

All of these techniques paved the way for Europe's most important innovation to sailing technology: the man-of-war. During the fourteenth and fifteenth centuries, Western Europe armed sailing vessels, which provided a key element to their emergence as the dominant world power.

A man-of-war was typically a three- or four-masted vessel rigged with square sails for long-distance travel. Lining each side of the vessel, either on deck or on a lower deck, were between six and ten cannons. The strategy for naval warfare, in which sailing ships fought one another, took shape around the man-of-war's layout and its ability to manage its source of power—the wind. Typically, enemy ships had to manage wind patterns in order to move to close quarters with one of its broadsides facing the enemy's ship. If maneuvered correctly, the ship could then fire its cannons at the enemy.

Henry the Navigator Exports Nautical Know-how

Prince Henry the Navigator of Portugal was a royal prince, soldier, and patron of explorers. Henry sent many sailing expeditions down Africa's western coast, but he did not go on them himself. Thanks to Prince Henry's patronage, in the mid-1400s Portuguese ships sailed to the Madeira Islands, rounded Cape Bojador, sailed to Cape Blanc, sailed around Cape Verde, and went as far as the Gambia River and Cape Palmas.

These expeditions were sent to create much-needed maps of the West African coast and, in the process, to defeat the dominance of Islam. In the process of spreading Christianity, the Portuguese sailors established new trade routes. Prince Henry helped begin the Great Age of Discovery that lasted from the 1400s until the early 1500s.

In order to disseminate new navigational practices and seafaring, Prince Henry started the first school for oceanic navigation in 1418. His school trained sailors in astronomy, navigation, mapmaking, and science. Their goal was to sail down the west coast of Africa.

Although his ships set out courageously, none of the early voyagers dared to go beyond Cape Bojador, a tiny cape somewhat south of the Canaries. They feared that beyond the Cape the sea became so shallow and the currents so unpredictable that no ship would ever return. From 1424 to 1434 Henry sent out fifteen expeditions, all of which did not dare to pass the Cape. Then, the squire Gil Eannes set out after promising Henry that he would attempt the task, and he succeeded. To avoid the shallows near the Cape, he sailed westward into the open sea, and when he turned east again, he found himself on the south side of the Cape. Finally, Henry's explorers had broken this important psychological barrier!

By the mid-1400s, Henry's voyages began to bear economic fruit. Slaves and seals, and later other trade articles as well, began to be sailed from the African coast to Portugal. By the time Prince Henry died, in 1460, the Portuguese had reached Cape Palmas (Liberia), and a trading post had been established in Arguim (an island near Cape Verde). His persistence had unlocked trade with the Muslim world.

The Age of Exploration from Europe

The boundaries of the known world now seemed to expand exponentially with each passing year. In 1492, a trip to the East, made by sailing westward around the world, brought Christopher Columbus to the New World. Although the area that he unlocked became known as the Americas, Columbus had originally set out to find an all-water route to the East Indies. Upon spotting the Americas, Columbus believed he had reached his intended destination. His oversight would not be learned for a decade. His naming of the Native peoples as "Indians" remains to this day.

Spanish explorers exuberantly followed Columbus to the Americas throughout the 1500s. The explorers sought to spread Christianity and to find resources of value, specifically gold and silver to be traded and to be used as currency for trading. Spaniard Hernando Cortez discovered an abundance of gold among the Aztecs in what is now known as Mexico. Stories of more gold to be found led him and other Spanish explorers to conquer most of Mexico and Latin America.

The discovery of silver soon also led to mining efforts in Mexico and South America. Most of this silver was traded with China, which used it as its primary currency. Explorers in the New World also found new products to trade and grow, including corn, tomatoes, tobacco, and chocolate. In some areas, these and other crops were grown on farms called plantations that were run and owned by Europeans but worked by natives or imported labor. Such plantations took advantage of the non-European climates to grow exotic products to be sold in European markets.

The push to explore and to eventually colonize the Americas was fueled by accounts written by many of the earliest explorers. Often through exaggeration and outright lying, the accounts of explorers were primarily intended to satisfy financial backers in Europe and also to stir others to begin their own ventures. This tradition, of course, begins with Columbus.

Columbus's Logic, 1492

In this passage from Columbus's narrative, one sees his effort to demonstrate the splendor of the New World in order to continue the financial support of his backers and to stimulate further exploration:

Friday, 19 October. Within three hours we descried an island to the east toward which we directed our course, and arrived all three, before noon, at the northern extremity, where a rocky islet and reef extend toward the North, with another between them and the main island. The Indians on board the ships called this island Saomete. I named it Isabela. It lies westerly from the island of Fernandina, and the coast extends from the islet twelve leagues, west, to a cape which I called Cabo Hermoso, it being a beautiful, round headland with a bold shore free from shoals. Part of the shore is rocky, but the rest of it, like most of the coast here, a sandy beach. Here we anchored till morning. This island is the most beautiful that I have yet seen, the trees in great number, flourishing and lofty; the land is higher than the other islands, and exhibits an eminence, which though it cannot be called a mountain, yet adds a beauty to its appearance, and gives an indication of streams of water in the interior. From this part toward the northeast is an extensive bay with many large and thick groves. . . . This is so beautiful a place, as well as the neighboring regions, that I know not in which course to proceed first; my eyes are never tired with viewing such delightful verdure, and of a species so new and dissimilar to that of our country, and I have no doubt there are trees and herbs here which would be of great value in Spain, as dyeing materials, medicine, spicery, etc., but I am mortified that I have no acquaintance with them. Upon our arrival here we experienced the most sweet and delightful odor from the flowers or trees of the island. . . . I am not solicitous to examine particularly everything here, which indeed could not be done in fifty years, because my desire is to make all possible discoveries, and return to your Highnesses, if it please our Lord, in April. But in truth, should I meet with gold or spices in great quantity, I shall remain till I collect as much as possible, and for this purpose I am proceeding solely in quest of them.[8]

Extending Trade Networks Through Colonies

Numerous wars at home did not prevent the new states from exploring and conquering wide portions of the world, particularly in Asia and in the newly discovered Americas. In the fifteenth century, Portugal led the way in geographical exploration, followed by Spain in the early sixteenth century. These nations were the first to set up colonies in South America and trading stations on the shores of Africa and Asia. They were soon followed by France, England, and the Netherlands.

In an effort to formalize the system of sailing, a colony was a fairly permanent tool for harvesting resources that existed in another place or climate. Typically, residents of the supporting nation populated the colony with military assistance. Each colonial enterprise, though, was founded with capital from its mother country, from either investors or kings and queens.

Each European nation utilized different methods of settlement and development. Spanish explorers eventually gave way to military and agricultural settlements, including pueblos and missions. French explorers established trading posts. British mariners, though, eventually emphasized colonies that might become permanent settlements. Often, these efforts were supported by joint-stock companies that combined investors and royal support from a nation to develop or harvest the resources wherever they occurred, such as VOC discussed above.

As overseas holdings increased and navigational capabilities grew, many European countries moved the armed trading idea to a more formal level by establishing national or royal navies. Clearly, control of the sea represented a direct correlation to economic development and stability. As more and more goods were moved by ship throughout evolving networks of trade, navies also became critical for protecting shipping channels and specific vessels. One of those dangers was yet another product of the international commons of the ocean: piracy.

EXPLORATION AND EXPANSION: GUNPOWDER AND CANNONS

Within the larger structure of managing wind power, the application of a separate technology could have enough of an impact to alter society. For instance, gunpowder may qualify, exactly, as a prime mover; however, its inclusion on sailing ships altered the entire system. Experiments in China and the Arab world perfected an explosive powder before the year 1000. Primarily potassium nitrate, it was used as an ingredient in metallurgical operations and in ceremonial fires. The Mongols acquired the technology and "weaponized" it by using it in bombards, a very early cannon, that were employed in sieges. The bombards, which would later become known as the cannon, helped Mongols conquer Chinese cities.

Then the technology was brought to Europe where the cannon helped to determine political power for the following few centuries. The prevalence of warfare during these years made any development of machinery for war of the utmost interest to royalty. The cannon, first used in battle in 1346, was massive and difficult to move. Despite the logistical difficulties in using them, cannons were used to help defeat castle-based nobility and, by 1500, the more successful centralized states were "gunpowder empires," including the Ottoman and emerging Russian empires as well as the European monarchies.

Although cannons were used extensively by Europeans against each other, they also became a primary tool for international trade once they were included on ships. Armed trade had been adopted by most European powers by the 1400s as they tried to break into existing Asian and Indian trade networks as well as to open new ones into the Atlantic and Pacific Oceans. Possibly the most famous example of using the cannon to help open new trade was Portugal, which, under the direction of Henry the Navigator, used armed navigation between 1415 and 1460 to open trade networks along the African coast. Armed traders including Columbus, Cortés, and others, would then travel to new parts of the world in the knowledge that they possessed the world's most potent weapon.

Whether used against enemies or to intimidate potential friends, the cannon traveled with most European explorers after 1450. In addition, the cannon became a crucial tool in Europe's perpetual wars during the coming centuries. As monarchies fought against one another in Europe, a weapon such as the cannon could completely alter the fate of a nation.

The Wheel Lock and Changes in Weaponry

During the Renaissance there were significant developments in the technology of weaponry, which gave added power to the armies of European monarchs or governments and helped them to increase their domains and trade networks. The primary change in the machines of war in the 1500s was their size.

The predecessor of the handheld gun, the wheel lock was the first self-igniting firearm. Developed in the early sixteenth century, it resembled a large version of a modern-day cigarette lighter. The wheel lock pressed a spring-loaded metal wheel against a sparking material, usually a small piece of iron pyrite. When the trigger was pulled, the tightly wound wheel was released and would spin rapidly, scraping against the pyrite. This generated sparks that fell into the priming pan, igniting its powder and then the main charge. Each firing of the gun required a recovery process referred to as "spanning the lock," in which the main spring was rewound using a special lever. The sparking material lasted for approximately a dozen shots before requiring replacement. The wheel lock took around a minute to load, prepare, and fire.

Historians for many years credited the wheel lock to the German Johann Kiefuss in 1517; however, drawings by Leonardo da Vinci dated to the early 1500s appear to show a wheel lock mechanism. The wheel lock was used along with the matchlock until both were replaced by the faster, simpler, and less costly flintlock of the 1600s. The matchlock used a lighted wick, whereas the flintlock could fire using a spark from a shard of flint stone. However, the ability of the wheel lock to fire reliably in wet conditions meant that well-made guns of this design would continue in use until the eighteenth century.

The Spanish Armada Expands the Militarization of the Sea

Combining the technology and know-how into one enterprise, the possibility of maritime dominance was first realized by the Spanish fleet, which became known as the Armada. However, throughout the 1500s and 1600s, the seas became one of the primary battlefields on which European monarchies fought for domination, power, and wealth at home. The greatest of these trade wars played out between Spain and England.

As early as 1585, Philip II, King of Spain, had begun to prepare a great fleet that, under the Spanish commander Santa Cruz, moved ongoing disagreements between the nations to sea—with ships from each nation privateering and harassing those from the other nation.

Finally, the Armada became the centerpiece of Philip II's plan to invade England, which was ruled by Queen Elizabeth. Aware of such plans, Queen Elizabeth set about raising funds to ensure that when the Spanish fleet came, England would be prepared. The great fleet left Spain in May 1588.

The Spanish fleet, which consisted of over a hundred ships, intended to sail into the English Channel. In the channel, the Armada expected to meet with the forces of the Duke of Parma, Philip's nephew, and then to continue as one force toward England. It was believed, then, that this force would overwhelm the English and defeat Queen Elizabeth.

In the English Channel, though, the Spanish suffered a humiliating defeat. The weather was dreadful. Suffering under the wind and rain, they were not able to compete with the superior English ships and war tactics. When the Armada fled, they had to follow a perilous journey around the coast of Scotland. Many of the Spanish ships were lost on the return trip. Overall, the Armada's effort to dominate the seas had failed miserably.

Although King Philip sent other fleets against England in the 1590s, none was as threatening as that of the great Armada of 1588. Most important, the English victory broke the era of Spain's maritime dominance and began an era of expansion for England, powered by its own control of the sea.

THE ATLANTIC SYSTEM INITIATES
THE GREAT REVERSAL

In the logic of economic expansion that drove European powers abroad, New World resources became sources of military and political power at home. Efforts to collect and develop these resources allowed many leaders and merchants between 1400 and 1800 to suspend consideration of the rights of Native peoples or of the human rights of laborers. By far, though, the most glaring example of this was the traders' ability to see humans—specifically Africans—as another commodity of value to be moved, sold, and collected.

Africans had been traded as slaves for centuries—reaching Europe via the Islamic-run, trans-Saharan, trade routes. Between 1450 and the end of the nineteenth century, slaves were obtained from along the west coast of Africa with the full and active cooperation of African kings and merchants. In return, the African kings and merchants received various trade goods including beads, cowrie shells (a type of money), textiles, brandy, horses, and guns.

By the mid-1400s, Portugal had a monopoly on the export of slaves from Africa. It is estimated that during the four and a half centuries of the transatlantic slave trade, Portugal was responsible for transporting over four and a half million Africans (roughly 40 percent of the total). However, during the eighteenth century when over six million Africans were traded, Britain was the worst transgressor—estimated to have been responsible for trading almost two and a half million Africans.

The system of trade that took shape in the Atlantic Ocean is often referred to as the "triangle trade." This term derives from a basic movement that currents and goods often followed. In this self-contained system, the eastward wind pattern, which blows on the southern part, came to be known as the "trade winds" because they enabled ships to cross the Atlantic. The westward wind pattern, blowing on the northern part, came to be known as the "westerlies."

Sailing ships were highly constrained by dominant winds, and, therefore, the trading system followed this pattern. Manufactured commodities were exported from Europe to go in two directions: toward the African colonial centers and toward the American colonies. Slaves then left Africa bound primarily for South American colonies (Brazil, West Indies). Tropical commodities (sugar, molasses) flowed from these colonies to the American colonies or to Europe. Within this Atlantic system, North America also exported tobacco, furs, indigo, and lumber to Europe.

CURRENT: Mercantilism and the System of Global Trade

The buying and selling of slaves marked the low point of the basic rationalization of this new European trading system. Just as Europeans reinterpreted resources anywhere as commodities at home, placing a value on the human body for its ability to perform labor (not paid to the worker) made economic sense because it could be profitable within the existing economic system.

From the markets and fairs of Europe, the interest in traded goods took form in the 1300s and fueled philosophical shifts in Europe that included an interest in individual profit. Eventually, Adam Smith, John Locke, and others would shape the liberation of capital from royalty into an economic system today called capitalism. The priority of this system was the individual's opportunity to derive economic benefit and profit.

Combining this cultural and social desire with advancements in ocean navigation made the European worldview one shaped by profits and commodities. Eventually, this worldview became known as mercantilism, which is described as economic nationalism for the purpose of building a wealthy and powerful state.[9] In 1776 Adam Smith's *The Wealth of Nations* coined the term *mercantile system* to describe the system of political economy that sought to enrich the country by restraining imports and encouraging exports.

This system dominated Western European economic thought and policies from the sixteenth to the late eighteenth century. The goal of these policies was to achieve a "favorable" balance of trade that would bring gold and silver into the country. In contrast to the previous agricultural system, the mercantile system served the interests of merchants and producers such as the British East India Company, whose activities were protected or encouraged by the state.

During the mercantilist period, military conflict between nation-states was both more frequent and more extensive than at any time in history. The armies and navies of the main protagonists were no longer temporary forces raised to address a specific threat or objective, but now were full-time professional forces. Their job was to assist and protect economic interests. Therefore, each government's primary economic objective was to command a sufficient quantity of hard currency to support a military that would deter attacks by other countries and aid its own territorial expansion.

The evolution of new economic ideas had blossomed through the new interactions with other nations and resources. The mercantilist system knitted together participating nations in a way never before seen.

MOVING ENERGY IN SHIPS: THE MIDDLE PASSAGE

No better image exists of the energy exchanges during the Age of Sail than the slave ship. Owned by parties vested in the systems that grew from the power of the wind, slave ships moved human energy to agricultural areas in need of labor. In completing its purpose, the slave ship participated in a system that perpetuated the reliance on human labor that reached back to the biological old regime. The new twist, however, came from the systematic application of commodification to human labor.

The leg of the trade system that carried Africans into slavery was often referred to as the "Middle Passage." European ships were loaded with groups of six people chained together with neck and foot shackles. On board, they were put below the decks, placed head to foot, still chained in long rows. Most Africans suffered from seasickness and additionally lost hydration due to vomiting. Additionally, poor food and fearful conditions contributed to diarrhea. The ensuing conditions below deck, of course, led to the outbreak of disease, including typhoid fever, measles, yellow fever, and smallpox.

The unhealthy conditions were made worse by the common practice of overcrowding a ship in order to maximize profit. The longer the ship was at sea the higher the slave mortality rate. Shorter voyages were expected to result in a 5–10 percent mortality rate; on longer voyages, though, traders expected between 30 and 50 percent of the slaves to perish.

The extreme human degradation of the Middle Passage exerted severe psychological shock on nearly every African. This was compounded by a common fear among the Africans that they had been taken by the Europeans to be eaten, to be made into oil or gunpowder, or that their blood was to be used to dye the red flags of Spanish ships. Of course, traders were instead attracted to the Africans' abilities as agricultural laborers and their adaptability to tropical climates. They were a source of human power that could be forcibly delivered to climates in which they would be forced to contribute to plantation enterprises. The system, literally, viewed their bodies as cogs in an agricultural machine.

The few remaining accounts of the Atlantic system, such as the narrative of Olaudah Equiano, remind us of the humanity that this system conveniently overlooked:

> At last, when the ship we were in had got in all her cargo, they made ready with many fearful noises, and we were all put under deck, so that we could not see how they managed the vessel. But this disappointment was the least of my sorrow. The stench of the hold while we were on the coast was so intolerably loathsome, that it was dangerous to remain there for any time, and some of us had been permitted to stay on the deck for the fresh air; but now that the whole ship's cargo were confined together, it became absolutely pestilential. The closeness of the place, and the heat of the climate, added to the number in the ship, which was so crowded that each had scarcely room to turn himself, almost suffocated us. This produced copious perspirations, so that the air soon became unfit for respiration, from a variety of loathsome smells, and brought on a sickness among the slaves, of which many died—thus falling victims to the improvident avarice, as I may call it, of their purchasers. This wretched situation was again aggravated by the galling of the chains, now became insupportable; and the filth of the necessary tubs, into which the children often fell, and were almost suffocated. The shrieks of the women, and the groans of the dying, rendered the whole a scene of horror almost inconceivable. Happily perhaps, for myself, I was soon reduced so low here that it was thought necessary to keep me almost always on deck; and from my extreme youth I was not put in fetters. In this situation I expected every hour to share the fate of my companions, some of whom were almost daily brought upon deck at the point of death, which I began to hope would soon put an end to my miseries. Often did I think many of the inhabitants of the deep much more happy than myself. I envied them the freedom they enjoyed, and as often wished I could change my condition for theirs. Every circumstance I met with served only to render my state more painful, and heightened my apprehensions, and my opinion of the cruelty of the whites.[10]

Similar to colonies, plantations and the slaves who worked on them allowed European powers to apply capital to co-opt the climate of far-off locations. Connected by sailing ships, the initial investment of European nations could then reap ongoing goods and revenue while growing sugar and rice in the Caribbean or Atlantic Ocean. Tied into this Atlantic trade system, the American South became an entire region based around plantation agriculture.

THE PERSISTENCE OF PLANTATION CULTURE IN THE AMERICAN SOUTH

Since the beginning of colonial settlement, agriculture in the southern United States was defined by an imbalance between population and land. Due to climate considerations and also the preferences of some of the initial European settlements, southern planters focused on crops such as tobacco, rice, cotton, and sugar. Each of these crops required large tracts of land as well as many laborers. For these reasons, many planters organized their land into plantations. These vast agricultural colonies presented a very different model of economic development.

By definition, plantations are large agricultural estates cultivated by bonded or slave labor under central direction. After being used on the islands of the Caribbean, slavery on plantations was introduced into North America in the British colonies of Virginia, the Carolinas, and Georgia. By the late seventeenth century, slavery was firmly established in Virginia and the Carolinas. Plantations were not the only agricultural type in any portion of the South, but it was widespread enough to cause many to identify it with the region. Geographer Sam Hilliard describes six essential elements to a plantation: generally over 250 acres in size; distinct division of labor and management; specialized production of one or two products or monoculture; located in the South, with a tradition of plantation agriculture; distinctive spatial organization to reflect centralized control; and particularly intensive use of human labor.[11]

Although most farmers in the antebellum South did not own slaves, those who did dominated agricultural production. Planters who owned slaves also possessed power, not just to dominate other human beings and profit from their labor, but also over the difficult environment. Slavery and exploitation of the environment went hand in hand.

One's mental picture of the plantation economy that dominated the antebellum South does not do justice to the diverse agriculture that dominated the subtropical climates of North America. Southern life and trade were largely organized by rivers. From the original tobacco sites along the James River in Virginia, growers of the "noxious weed" spread to the north, south, and west. Even as tobacco growers moved inland to the Piedmont Mountains, most plantations were located on or near rivers. By 1800, however, growers had expanded so far inland that about three-quarters of all tobacco was sold first within the United States before being sold at ports to companies that would transship to England.[12]

Historian Lewis Cecil Gray wrote that many southerners "bought land as they might buy a wagon—with the expectation of wearing it out."[13] Through their land-use practices and crop selection, many southerners mined the fertility out of the land. Typically, such farmers would then move farther inland to begin anew. Partly for this reason, the initial push westward came over the Appalachians from the mid-Atlantic and South. Geographer Terry Jordan refers to this early push as the "backwoods frontier," which was a settlement process that also established a fairly consistent cultural pattern.[14]

The infrastructure of the young United States did not function equally. Regional distinctions became apparent during the 1800s. Primarily, while the Northeast and Midwest moved more deeply toward new sources of energy and industrialization, the southern United States resisted infrastructural development and expanded the use of slaves in order to expand the growth of cotton through the South and to the Mississippi River. Although the scale and scope of southern agriculture grew massively, rivers remained the primary courses of trade. Historian Ted Steinberg writes: "There is nothing the least bit natural about slave labor, but in the antebellum South, at least, it owed its rise to a climate that favored the growth of short-staple cotton. The development of the Cotton Belt rested on a set of climatic conditions; without them it is hard to imagine slavery taking on the role that it did in southern political culture."[15]

Despite a similar westward migration, the development of cities and towns in the southern United States took a different path than that of the northern and western United States. As the Cotton Belt grew in the interior South, towns were created and population increased in existing towns. This growth, however, was slower than corresponding urban expansion in the North. A clear differentiation evolved between North and South in the nineteenth century. The influence of the planter culture on the growth of cities in the South and on the type of cities that did emerge is evident in both social and economic spheres.

In the nineteenth-century South, the plantation, not the city, was seen as the location of opportunity. Early on, in Virginia and the Piedmont, the plantations had little need of town services when the excellent river transportation facilitated trade between plantation and England with relative ease. Later, when American ports became primary destinations of planter production, each plantation formed a fairly self-contained unit of operation and life, separated from other plantations, often more closely tied to the port of destination of its crop than to local towns. Plantations generally produced many of the items that in other areas of the United States were available only in towns.

SEAPORTS AND HINTERLAND FORMALIZE THE ENERGY SYSTEM

As the Atlantic system radically altered the human approach to agriculture, the United States became a dramatic example of how it might also alter large-scale economic development. The familiar mechanisms and systems of earlier days expanded to become increasingly complex, capable, and even monstrous. As with European settlement itself, this growth began with the sea. When the United States was established in 1776, the nation structured itself around the nineteen eastern port cities that had established themselves during the colonial era.

Early on, colonists had clustered into settlements along the water for reasons of safety and trade. The early port settlements were lifelines connected back to Europe. Boston was founded in 1631 and Manhattan Island around 1625. Philadelphia followed in 1681. These early ports combined with southern ports on the Atlantic, including Savannah, Georgia, and Charleston, South Carolina, to provide the connection for trade to Europe. In each case, the developing ports centered cities on rows of wharves that grew out of tightly packed streets full of storefronts and warehouses.

Often, these seaports were discernible by the material used for street construction. Sailing ships relied on maintaining balance in stiff winds and high seas. Below the decks, most vessels carried stacks of stones, usually larger in size than a brick. Called ballast rock, this material could be shifted around as the ships' stores grew full or empty. When ships called on foreign ports, they often unloaded some of this rock or added additional ballast. Most ports wound up with a superfluous supply of ballast rock that could then be woven into making cobblestone streets. The collection of stone from around the world can make streets unlike any others in the United States. In addition to the ballast rock streets, products of the sea such as seafood, oils, and bone dominated port cities. More important for national development, though, ports or entrepôts fueled the development of specific "hinterlands." Relying on ports for trade in either direction, hinterlands grew around the idea of access—normally, this was defined as a wagon ride of no more than a few hours.

"Commerce, not shipbuilding or fishing," writes historian Benjamin Labaree, "is what distinguishes significant seaports from other seacoast communities."[16] He describes seaports as entrepôts for culture and goods. As such, the entranceways also acted in reverse: they served broad swaths of interior land that could access the port city. Referred to as hinterland, these interior areas had a symbiotic relationship with seaports thanks to the economic possibilities that the ports embodied.

Of North America's coastal towns, New York was the most favored by nature to become a major seaport. Early on, however, New York lost its initial lead to Philadelphia, which became the nation's leading port by 1760. By 1815, all of America's population centers could be found on water. The rapidly growing port of Cincinnati on the Ohio River was the only population center not located on the Atlantic Ocean. Of the nation's population of eight and a half million inhabitants, roughly 85 percent lived along the Atlantic Coast, with about half of the nation's population in New England and the mid-Atlantic states. With roads largely undeveloped, Americans depended on the waters for food, transportation, and trade. The French observer Alexis de Tocqueville wrote in 1830:

No other nation in the world possesses vaster, deeper, or more secure ports for commerce than the Americans. . . . Consequently Europe is the market for America, as America is the market for Europe. And sea trade is as necessary to the inhabitants of the United States to bring their raw materials to our harbors as to bring our manufactures to them. . . . I cannot express my thoughts better than by saying that the Americans put something heroic into their way of trading.[17]

CONDUIT: The Wharf

Essential technologies tied the goods and materials being produced on land to the vessels arriving by sea. By the 1770s, U.S. boatyards constructed five hundred vessels per year. Wharves, in part a legacy of Great Britain, provided berths for deepwater vessels along the waterfront of every seaport. Historian John Stilgoe is careful to discern between maritime terms often used interchangeably. He writes: "A dock exists only as a space, an open area adjacent to a wharf, a pier, a landing stage, even a bit of marsh." However, continues Stilgoe, "wharf typically designates a structure set on wooden spiles and carrying a wooden deck, whereas pier designates a stone, earth-filled, or steel-and-concrete construction carrying a deck made of anything but wood."[18]

The capital for such improvements came from merchants acting together or individually. In such cases, wharves often bore the name of their financier. Shipowners also erected piers or other devices to mark obstructions and shoals that impeded navigation, but major harbor improvements, such as dredging and jetty construction, required the later assistance of the federal government. In 1716, America's first lighthouse was placed on Little Brewster Island near the entrance of Boston Harbor. Although a few sea captains owned English charts, local residents known as "coast pilots" came aboard incoming vessels to guide them with their knowledge of the local waters. Each of these practices created an infrastructure for maritime trade that kept the ocean as a primary influence in the lives of many Americans.

In places such as Boston, the wharves became the stepping-stone to converting ocean into land. "Where nature has provided some but not all of these geographical requirements," writes Labaree, "man has not hesitated to intervene."[19] Prioritizing access to the sea, Boston transformed itself after its founding in 1630. Reaching out to the sea, Bostonians had constructed seventy-eight wharves by 1710. The most impressive of these, Long Wharf, extended eight hundred feet from the shoreline. The competition of this new era spurred seaport and city growth in the 1800s. Boston, of course, had the natural benefit of open access to Atlantic trade. However, nature had not blessed Boston with a great deal of surface area. Bostonians demonstrated great resourcefulness by creating more land. The tradition began in the 1600s and continued through the twentieth century. By then, Boston had more than tripled its size!

The proximity of Boston, hotbed of the revolution, to the early New England settlements made it the nation's most active port through the 1700s. This activity, however, was more than matched by its furious efforts at filling. Because of this early filling, the three original hills of the city (Pemberton, Beacon, and Mount Vernon) no longer dominate the landscape. In addition to creating new land, these surrounding hills were carted to the sea's edge and dumped in the coves to provide more area for building. As Boston grew, it became the primary regional port, tying together a hinterland of smaller ports and fishing towns. New York, Philadelphia, and Baltimore, however, were primarily supported by agricultural regions inland.

FISHING THE ATLANTIC PROVIDES ABUNDANT PROTEIN

Goods passed over the sea, but any port was also supported by resources that came from the depths of the ocean that were accessed by sailing ships. Gloucester, Massachusetts, served as the great center for American fishing. Soon, the state of Maine led in the total tonnage of vessels devoted to fishing. By the early 1830s, fishermen ranged beyond the Grand Bank of Newfoundland to extend into Georges Bank, east of Cape Cod. This expansion proved very dangerous for fishermen. For instance, a single 1846 storm brought a loss of twelve vessels from Marblehead, Massachusetts. In 1851, Gloucester lost even more vessels to a large storm.

Fishing for cod and mackerel with handlines from the main vessel, which was usually a fairly large schooner, made up the largest portion of the New England fishery. Mackerel fishermen chased large schools of the fish, which stayed close to the surface of the water. Early in the 1800s, the mackerel fishery received a boost from the introduction of the shiny jig that served as a lure. In the 1850s, fishermen began using huge purse seine nets. These wide nets would be cast into the water and then drawn into a tight purse shape by pulling on a drawline. First used for mackerel, the purse seine was eventually used for other fish as well. The highly demanded mackerel was eaten fresh or salted, which allowed the fish to be preserved and saved for later use.[20]

The cod fishery also endured significant changes around the mid-1800s. Early cod fishermen used a series of sharp, baited hooks dangled from the side rail of a vessel. In the mid-1850s, New England fishermen adopted the European model of a trawl line. In this method, fishermen laid a long line on the bottom with many baited hooks attached at intervals. To set this gear, schooners began to carry flat-bottomed, nearly double-ended dories, which were launched from the fishing schooner each day. It is estimated that this new method was roughly three times more effective than the previous single-line method.

Another important innovation dealt more with fish storage. Around 1860, schooner captains began using ice to increase the market area for fresh fish. Although salted cod remained suitable for trade with even distant lands, icing fish increased the availability of various types of fresh fish along the eastern coast. This included shellfish, which had been largely depleted in many urban ports. Popular delicacies such as oysters could now be transported from other regions (such as the Chesapeake Bay). In addition to being eaten, these oysters also were used for some of the first American aquaculture: if shells were laid down in clutches at spawning time, they could collect juvenile oysters and replant them in other areas. Ice and the construction of transportable live-boxes also allowed lobster to grow more popular in the 1840s. Throughout the early 1800s, though, lobster and menhaden were used mainly as fertilizer. Boiled, pressed, and dried, farmers called the fertilizer "fish guano."[21]

The American fishery marked some of the earliest efforts to limit or regulate resource use. The earliest regulations attempted to balance the taking between Canadian and U.S. ports. In 1818, American and British negotiators agreed that American fishermen could set up processing sites at certain abandoned coasts along Labrador and Newfoundland. These agreements generally demanded that American fishermen pay a fee per ton. Such additional fees kept American fishermen at a competitive disadvantage. Regardless, using the power of the wind, the fishing fleet grew in size from 37,000 tons in 1815 to 163,000 by 1860.

A WHALE OF A BUSINESS

The energy at work in any harvest from the sea was drawn from the wind; however, the sea proved a gateway for a transition to other prime movers. The bounty of the sea could also be transformed into one of America's first full-blown industrial endeavors. Singular whale kills supported many early small seaports. In the eighteenth and nineteenth centuries, though, whaling evolved to be a major American industry to provide oil for lamps and spermaceti for candles. It also had a marked influence on the territorial expansion of the United States and the management of U.S. diplomatic relations.

Beginning as a whaleman on Nantucket, Joseph Rotch saw the potential for growth in whale oil. He confronted the limits being placed on suppliers by manufacturers of oil and candles by joining the competition. By manufacturing his own candles, Rotch became the oil industry's first magnate. Essentially, he had taken control of the single resource that offered Americans consistent light. The next stop in Rotch's empire of light came when his interest in manufacturing led him away from Nantucket and to New Bedford, a small village of recent creation in the mainland town of Dartmouth. Rotch bought ten acres along the Acushnet River in 1765 and moved his family in 1767. He only intended one thing for New Bedford: that it should serve as the world center of whaling, oil trade, and candle manufacturing—in other words, of light. In 1768, the first of New Bedford's candleworks opened. Rotch owned it with a Newport manufacturer, Isaac Howland.

Rotch used New Bedford to tie together his vast holdings, which grew in 1771 when his son opened the first candle manufacturing facility on Nantucket. While New Bedford was not a corporate entity, it functioned to bring organization to the whale oil business. The town linked a series of operations that included catching, processing, and distributing whale oil and its derivatives. After all the squabbling, Rotch had emerged in the most powerful position; it was fitting that in 1769 the United Company approached the New Bedford consortium in hopes of including them. Howland agreed, but Rotch remained dubious. Before the agreement could be pursued further, the American Revolution interceded. By 1775 British impressment from whaling ships had become so problematic that Rotch and the others ordered their ships off the sea. They would wait out the war that came next.

Many national observers had noted the great possibilities of New England's whale fishery; nations had even begun to compete for its practitioners. John Adams and Thomas Jefferson noted the industry's critical importance to the young nation at the close of the eighteenth century. As Jefferson considered the nation's trade difficulties in 1789, he focused specifically on the whale fishery and its role in relations with France and Britain. In doing so, Jefferson offered whale oil as second only to tobacco in its importance as an American export. Proving him correct, the ensuing decades marked a bidding war for the fishery and those on Nantucket Island who had perfected the pursuit.

Following the Revolutionary War, Britain realized the importance of the whale products now being harvested almost exclusively by its former subjects on Nantucket. Britain made extensive efforts to recruit the whalemen to become British subjects and operate from outposts in Halifax and Nova Scotia. France, in retaliation, made a similar effort to recruit the whalemen from Nantucket in 1785, establishing an outpost for them in Dunkirk.

The rigging on sailing ships varied considerably. In this case, the large brig is powered by numerous square sails. LIBRARY OF CONGRESS PRINTS AND PHOTOGRAPHS DIVISION

Jefferson deduced that "it would be safest in every event to offer some other alternative [to the whalemen]. . . . The obvious one was to open the ports of France to their oils, so that they might still exercise their fishery, remaining in their native country, and find a new market for its produce instead of that which they had lost."[22] Despite the U.S. emphasis on free commerce, Jefferson urged Congress to take action to combat these aggressive maneuvers by competing nations. Later in 1789, legislation was passed to moderate the prohibitions and monopolies being charged and offered by other nations and to stimulate the further development of the American whale fishery. Clearly, the significance of this pursuit far outweighed self-sufficient fishing.

The necessity of such negotiation demonstrates the global importance of the commodity in the late eighteenth century; however, it also suggests an ongoing source of conflict within the industry: while whalemen could be considered a skilled population, their quarry limited the industry's capability. The crude style of the hunt, in fact, placed "blubbermen" near the bottom rung of maritime occupations. The supply of whales could not magically increase without new technologies. Whales exist as a common resource with neither the control nor reliability of an owned or land-based resource.

In the case of whaling, ethical values control or guide the pursuit of energy. However, the intensity of the pursuit of energy was changing. Once whaling moved beyond the Nantucket model, the enterprise had become industrialized. New Bedford, the town replacing Nantucket as the fishery's hub, exemplified the instrumentalized state of whaling, moving it beyond maritime enterprises to encompass scale, scope, and organization of the industrial age. As the scale of the industry shifted in the early 1800s, variations in the whale population required the fleet to extend the hunt beyond the Atlantic. This extension clarified the need for a base seaport much larger and complex than Nantucket.

Whether the product was whale oil or china that had just arrived from a distant land, coastal development fueled internal developments that grew increasingly distant from the sea. New opportunities sprung up along the paths used to move these goods throughout the ports' hinterland.

CONDUIT: Lamp Illumination

The key land-based use that connected the pursuit of whales for oil was illumination, first in candles and subsequently in oil lamps. Sperm candles were first made in Newport, Rhode Island, in 1750, and throughout their use they remained a premium item, primarily exclusively available to wealthier consumers in Europe and the United States. During the late 1700s, the quest to democratize various aspects of higher standards of living drove American statesman and inventor Benjamin Franklin to focus his attention on perfecting the oil lamp. Specifically, Franklin in the late 1700s experimented with lamps that burned two flames, thus doubling their illuminative capabilities.

Following up on the work of Franklin, French inventors and particularly Aimé Argand, a physician and practical chemist in Geneva, Switzerland, focused on the lamp's updraft. In his design, Argand fitted a sheet-iron chimney over a metal tube extending through the base of the lamp to admit air from below and encased it with a circular wick. Eventually, he substituted this metal chimney with a glass one and extended the area of combustion without cutting off the light from the flame.[23] The result, wrote one journalist, was a "flame like the fire in a furnace" as opposed to common lamps that burned "like an open fire."[24] The distinction, of course, was the intensity of the flame and the brightness of the light that it dispersed.

Although efforts to perfect lamp design continued during the first decades of the 1800s, the additional revolution of lamp technology was its ability to accept a variety of fuel oils. This flexibility allowed the same or similar lamps to burn oil made from a variety of animal and vegetable fats, the effectiveness of each which Franklin and other scientists carefully evaluated. Such comparisons were made simpler in that each fuel could be burned in an identical lamp. Particularly in the American market, innovators experimented with variations of lard oil and also with distilling illumination oil from turpentine (which was derived from pine bark). By 1830, American Isaiah Jennings took out a patent for camphene, redistilled spirits of turpentine, which, burned alone or mixed with alcohol. Camphene became a leading synthetic oil lamp illuminant in the United States and would eventually be derived from petroleum and solid mineral bitumens, such as coal.[25] Whale oil lamps could be easily converted to burn camphene by the 1840s.

Harvesting a whale at sea was an extremely dangerous early energy exchange. LIBRARY OF CONGRESS PRINTS AND PHOTOGRAPHS DIVISION

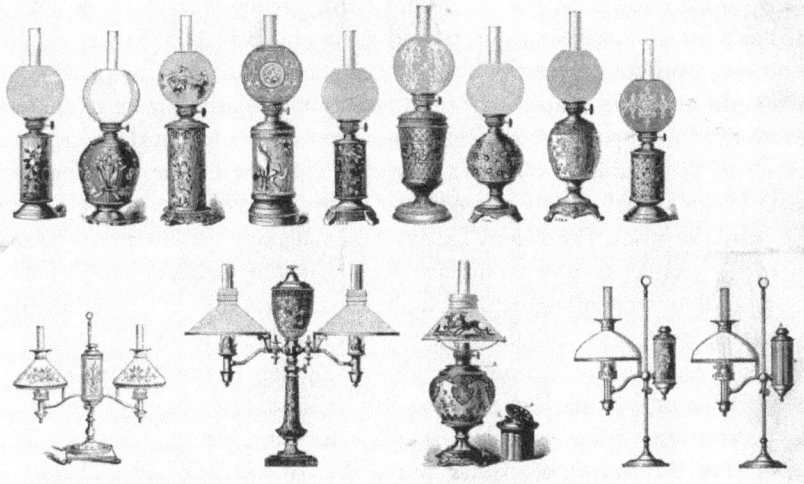

Animal fats such as the blubber from a whale could be hunted, harvested, and rendered for use in oil lamps. LIBRARY OF CONGRESS PRINTS AND PHOTOGRAPHS DIVISION

CONCLUSION: THE WAR OF 1812 AND AMERICA'S ENERGY TRANSITION

The economic potential of trade through the systemization of sailing ships allowed one of these colonies, the United States, to pursue independence. It makes sense that the new nation was established with trade—among other ideals—at its core and that its rebellion took root in its seaports, which harkened back to the earliest days of European settlement. The conflict that had begun with England had not been settled with the Revolution of 1776. Especially in matters relating to the natural resources of North America and their shipment to Europe, great animosity still swirled around trade disputes between the nations during the early 1800s.

The War of 1812, which pitted the United States against Great Britain, occurred when these disagreements escalated. The war started in 1812 and ended in stalemate in 1815. The root of the conflict concerned the rights of American sailors who were being impressed to serve in the British Navy. The major military initiative of Britain during the war, though, was more related to trade: the British blockade of ports such as Philadelphia nearly crumbled the economy of the young republic. Although the British fleet carried out attacks on Yorktown and along the Great Lakes, they clearly set their sights on the port of Baltimore.

The British crown sought to capture Baltimore by way of combined land and naval attack. Neither front proved successful. On September 12, 1814, Baltimore troops fought for two hours to delay the British land forces before they reached the city. The attacks on Baltimore, however, derived from its economic significance in the global Age of Sail. The port benefited from the construction of internal trade routes, including roadways and two canals: the Chesapeake and Delaware Canal, which opened in 1829 to link the bay with the Delaware River, and the Susquehanna and Tidewater Canal along the lower Susquehanna, which diverted Pennsylvania produce away from Philadelphia. The benefits of this expanding trade infrastructure should remain with the United States, argued Americans; by contrast, Britain made one last attempt to retain North America's economic role as one of its most profitable colonies in the Atlantic system of trade. The moment marked an inevitable crisis of the Atlantic system as it marched forward; however, it also ushered in a clear transition in energy usage.

Ironically, the British blockades of the War of 1812 helped to launch the United States into its next economic era: moving the United States more swiftly toward its industrial future. Depleting fuelwood supplies, particularly around Philadelphia, combined with the British blockade to create domestic interest in using the anthracite or hard coal made popular in Britain (discussed in chapter 3). Historian Martin Melosi writes, "When war broke out . . . [Philadelphia] faced a critical fuel shortage. Residents in the anthracite region of northeastern Pennsylvania had used local hard coal before the war, but Philadelphia depended on bituminous coal from Virginia and Great Britain."[26] Coal prices soared by over 200 percent by April 1813. Philadelphia's artisans and craftsmen responded by establishing the Mutual Assistance Coal Company to seek other sources. Anthracite soon arrived from the Wilkes Barre, Pennsylvania, area. After the war, industrial use of hard coal continued to increase slowly until 1830. Between 1830 and 1850, the use of anthracite coal increased by 1,000 percent.[27] The transition to fossil-fueled energy that had begun in Europe and Britain was now poised to define the emergent, independent nation. And, ultimately, with so little existing infrastructure, the United States might industrialize much more rapidly than could its competitors.

NOTES

1. Marcus Rediker, *Slave Ship* (New York: Penguin 2008), p. 50.

2. Alfred Crosby, *Children of the Sun* (New York: Norton, 2006), p. 48.

3. Gavin Menzies, *1421: The Year China Discovered America* (New York: William Morrow, 2003), pp. 78–80.

4. Menzies, pp. 80–81.

5. Menzies, pp. 79–81.

6. Robert Marks, *The Origins of the Modern World* (Lanham, MD: Rowman & Littlefield 2019), pp. 42–44.

7. Menzies, pp. 66–70.

8. "The *Journal* of Christopher Columbus (1492)," The History Guide, https://www.humanities.uci.edu/mclark/COURSES/EAL/E102BW2010/ColumbusJournal(1492).htm. Accessed on January 6, 2022.

9. Marks, pp. 92–93.

10. Gustavus Vassa, "The Life of Olaudah Equiano," https://www.litcharts.com/lit/the-life-of-olaudah-equiano/chapter-2. Accessed on January 6, 2022.

11. Michael Conzen, *The Making of the American Landscape* (New York: Routledge 2010), p. 106.

12. Conzen, p. 110.

13. Theodore Steinberg, *Down to Earth* (New York: Oxford, 2018), p. 74.

14. Terry Jordan, *North American Cattle-Ranching Frontiers* (Albuquerque: University of New Mexico Press 1993), pp. 2–3.

15. Steinberg, p. 87.

16. Benjamin Labaree et al., *America and the Sea* (Mystic, CT: Mystic Seaport Museum 1998), p. 23.

17. Alexis de Tocqueville, *Democracy in America* (New York: Library of America, 2004), pp. 400–407.

18. John Stilgoe, *Alongshore* (New Haven, CT: Yale University Press 1994), pp. 208–211.

19. Labaree, p. 10.

20. Labaree, pp. 258–259.

21. Labaree, p. 265.

22. Thomas Jefferson, "Observations on the Whale-Fishery," *Public and Private Papers* (New York: Vintage Books, 1990), pp. 382–383.

23. Harold F. Williamson et al., *The American Petroleum Industry* (Chicago: Northwestern University Press 1963), pp. 30–31.

24. Williamson et al., p. 32.

25. Williamson et al., p. 33.

26. Martin Melosi, *Coping with Abundance* (Philadelphia: Temple University Press, 1985), p. 84.

27. Martin Melosi, *The Sanitary City* (Pittsburgh: University of Pittsburgh Press, 2008), pp. 26–27.

TRANSITIONING BY THE NUMBERS
Industrialization

Industrialization includes the era of proto-industry in which new energy exchanges occurred singularly. The Industrial Revolution that grew out of Europe during the early 1700s was largely a transition in the use of technologies and systems that extended and expanded the capabilities of humans. These systems drew from the harvest of the massive reservoir of fossilized energy that had formed in Earth's geology over millennia.

Coal use clearly defined early industrialization globally.

Per Capita Coal Production, 1800s–1910 (TJ per 1,000 inhabitants)

	1800	1850	1880	1910
Africa				1,374
Americas	109	3,717	20,870	76,145
Asia	0	0	5.5	975
Europe	1,493	7,139	20,414	35,582
Oceania	0	0	9,822	41,082
Rest of World	321	1,758	6,475	17,887

SOURCES: From Warde, *Power to the People*, p. 140. Etemad and Luciani, 1991, 202.

Per Capita Coal Consumption, 1815–1931 (GJ per year)

	England	Germany	France	Netherlands	Sweden	Portugal	Italy	Spain
1815	48.5	1.5	0.9	1.2	0.2			
1840	62.2	3	3.5	3.8	0.3			
1870	112.4	20.4	14.8	14.8	3.2	1.2	1.2	2
1890	136.5	45.6	27.6	24	10.2	3.3	4.5	5.2
1913	135	86.8	46.8	51.8	29.3	6.1	9.6	11.6

SOURCES: From Warde, *Power to the People*, p. 140. See www.energyhistory.org.

Making coal into productive power for various uses began with its transformation into steam.

Stationary Steam Power Availability in Europe and United States in 1840 and 1870 (hp per 1,000 people)

	c. 1840	c. 1870
Britain	10.5	81.7
France	1.6	15.5
Prussia	0.4	15.9
Belgium	6.1	34.6
Sweden	0.3	1.8
Czechia	0.1	7.5
USA	2.3	37.1

SOURCES: From Warde et al., p. 184. Allen, 2009, 179; Warde, 2007, 75; Landes, 1969, 221; Kander, 2002; Myska, 1996, 255. The population of Czechia is assumed to be 70 percent of that supplied by Maddison for Czechoslovakia.

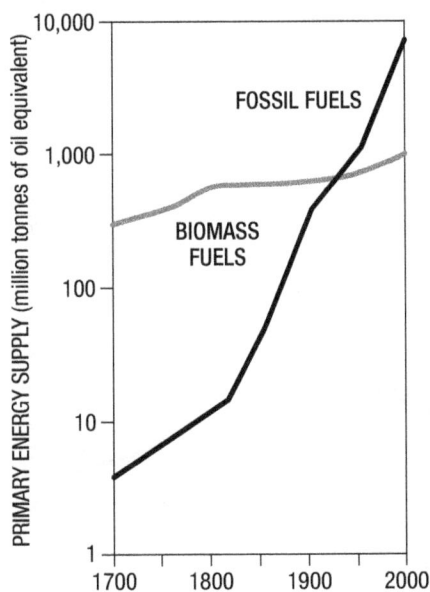

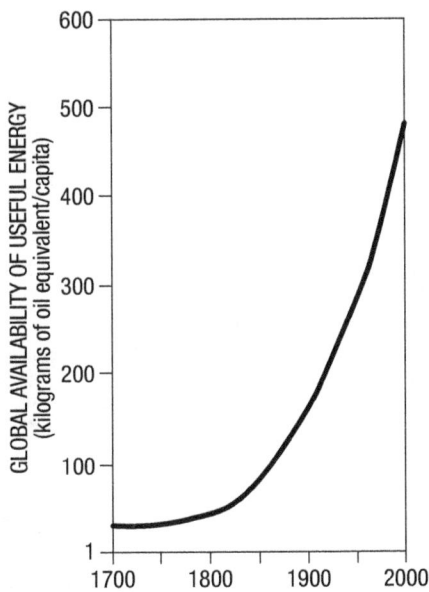

Biomass versus Fossil Fuels. ASTRID KANDER, PAOLO MALANIMA, AND PAUL WARDE, *POWER TO THE PEOPLE: ENERGY IN EUROPE OVER THE LAST FIVE CENTURIES* (PRINCETON, NJ: PRINCETON UNIVERSITY PRESS, 2014).

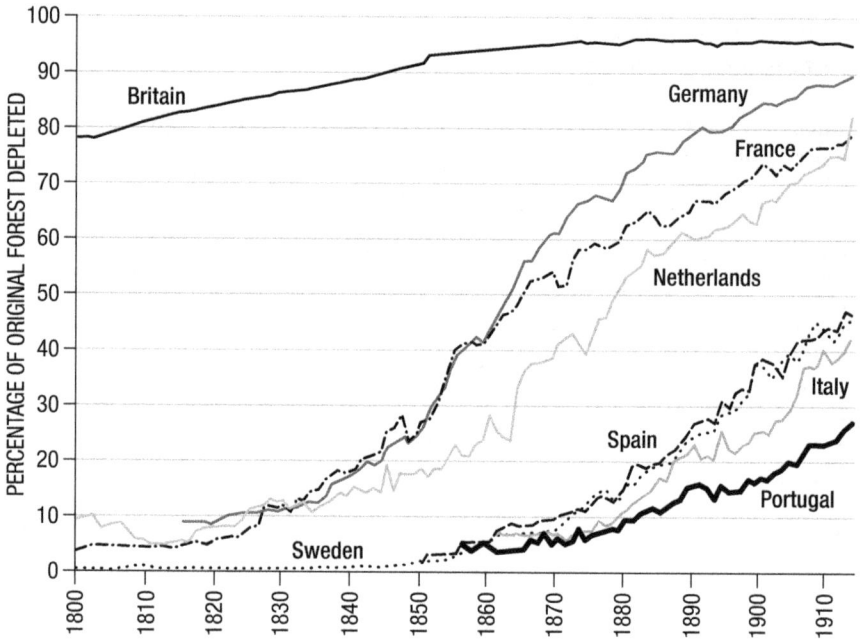

European Forest Depletion. ASTRID KANDER, PAOLO MALANIMA, AND PAUL WARDE, *POWER TO THE PEOPLE: ENERGY IN EUROPE OVER THE LAST FIVE CENTURIES* (PRINCETON, NJ: PRINCETON UNIVERSITY PRESS, 2014).

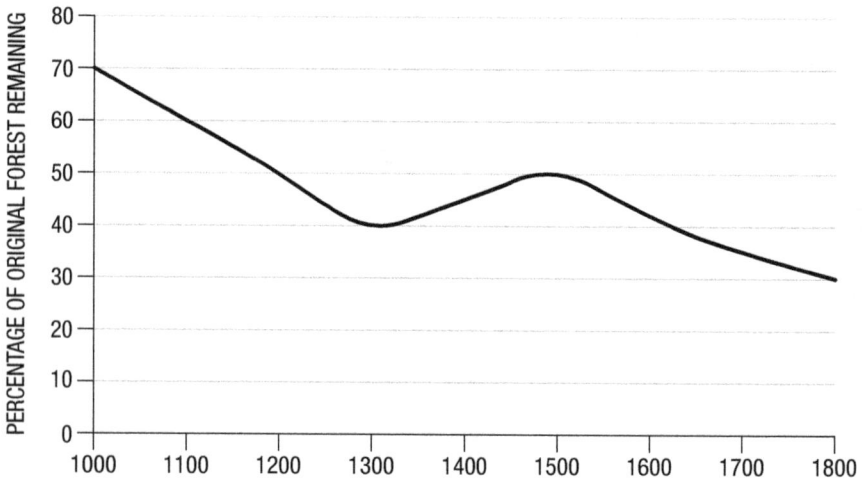

Forest Depletion in Europe with the Transition to Coal. VACLAV SMIL, *ENERGY AND CIVILIZATION: A HISTORY* (CAMBRIDGE: MIT PRESS, 2017).

3

Fossil Fuels and the Transformation of Human Work

GROUND: LEPIDODENDRON

Against the tropical backdrop of locations such as Greenland and Newfoundland at that time, huge Lepidodendron trees could grow to 175 feet tall, and they towered as a vivid demonstration of the success that was possible if photosynthesis fully carried out the energy of the Sun. The name is Greek for the word scale *because of the beautiful lizard-skin bark that covered the six-foot girth of these monstrous trees. As the trunk narrowed toward the tree's distant top, its broad branches shortened in length and dwindled in number. From them came narrow leaves up to a yard long that gave the tree the look of a shaggy, green pole.*

In other parts of the world such as Britain, similar trees grew in the tangled jungles of the Carboniferous period (360 to 290 million years ago) that were known as Sigillaria. Paleobotany texts described it more "like a huge barrel than a full-fledged tree." Historian Barbara Freese writes: "And then there were the ferns, primitive and highly successful plants, botanically related to the ones you may have growing in a pot in your home—except they had trunks and were thirty feet high."[1]

In total, this enormous global jungle held an unimaginable mass of vegetation. Over millions of years, they failed to decay the way that more moderately sized plants do. When a plant normally dies, oxygen penetrates its cells and decomposes it into carbon dioxide and water; by contrast, as the dense Carboniferous plants died, they often fell into oxygen-poor water or mud where they were covered by other decaying plants in a very slow, elongated burial process. As seas rose over this decaying material and then fell back, oxygen could not reach the buried plants to complete the decomposition work. First becoming peat, the spongy mass of carbon-rich plant material was slowly squeezed in a slow-cooking process that eventually hardened the peat into coal. As Freese writes: "It wasn't just the forest's carbon that ended up trapped in the coal, but the energy it had accumulated from the sun over millions of years. Instead of dissipating with the plants' decay, that energy was tucked away into the dark recesses of the earth, at least until the amphibians crawling across the forest floor evolved into creatures capable of digging it up."[2]

In the cases of Lepidodendron and Sigillaria, in particular, the mineral that they formed took shape beneath a living community of humans that would steadily move through patterns of life based in the biological old regime until they began to organize themselves for more even as they sailed, traded, and farmed. By the time these efforts took full fruition, Newcastle, England, served as the prime producer of the coal to drive humans into the industrial era.

Although the power of the wind proved to be a significant transition in the development of nations, the economic advancement of European powers enabled the flexibility for many of these same nations to consolidate resources and to seize new technologies. Often, these innovations continued to rely on the network of connections provided by wind's power. Early on, industrial endeavors grew out of small-scale efforts that often fell within overall patterns of the biological old regime.

What historians of technology refer to as the great transition is not necessarily confined to the emergence of the Industrial Revolution in the mid-1700s. In order to reach that revolution, a "great transition" was necessary in intellectual thought and, in addition, the availability of energy resources. Biomass fuels such as wood and charcoal had been in use for centuries, but they could not necessarily support an entirely new infrastructural system of machines. Coal, however, emerged as a prime mover during the 1600s and fulfilled the essential requirements of the new system.

Similar to some other transitions, supply proved to be a primary catalyst. After England experienced serious shortages of wood in the 1500s, domestic coal extraction became the obvious alternative. Most of the existing coalfields in England were opened between 1540 and 1640. By 1650 annual coal output exceeded two million tons. Acquired through improvements in mining, coal use would rise to ten million tons by the end of the 1700s.[3]

Coal itself possessed power possibilities far in excess of previous resources and made new industrial capabilities possible. Primary among these was the steam engine. These related innovations built on this well-known prime mover to create an instrumental transition of such significance that it has long been known as the Industrial Revolution, the period in human history that rises to the level of an intrinsic shift in energy use that rivaled that of the agricultural revolution that preceded it. However, there was no particular eureka moment to ensure that humans learned to burn fossil fuels in order to release stored energy. Instead, there was a steady stream of innovations teamed with a consistently growing

Coal supplies were enhanced by extensive mining by humans throughout the globe.
LIBRARY OF CONGRESS PRINTS AND PHOTOGRAPHS DIVISION

willingness to derive work and labor from stored reservoirs of energy instead of from actual animal and human power. Overall, the greatest single change brought by this shift was the release of human accomplishment from the natural limits of those bodies.

Through its depiction of humans' integration of coal power, this chapter emphasizes the shift in the human condition when energy of any type emerged as a most crucial component to human economic production.[4] Previous stages of labor were limited because their prime mover was as well: whether because of rays of the Sun, gusts of wind, the pull of a human or animal muscle, or the flow of a river, human productivity suffered under the governor of nature's innate limits. By the close of the twentieth century, the standard paradigm of energy use derived from the burning of hydrocarbons and nature's limits no longer constricted human capabilities. This way of harvesting power will be so accepted and normal that, at that juncture in the twentieth century, other methods of obtaining energy—even if they have been used for thousands of years prior—will be classified as "alternatives." Burning fossil fuels was the new normal, acceptable. Today, as we return to some of these alternative methods for energy production, we place the era of industrialization within a critical additional reality of historical context: the expansion of the last 250 years that was made possible by the burning of fossil fuels has contributed to—if not entirely stirred—climate change in the twenty-first century.

THE INTELLECTUAL UNDERPINNINGS
OF THE MACHINE

Examples of proto-industry were evident in various pockets around the English countryside by the 1300s. Discussed in chapter 2, the prime movers providing motive power to factories included wind, water, and tides, each one very limited and tied to specific locations. Rapidly, these early industries made flexibility a valuable commodity and increased the potential of undertakings that did not rely on static geographical features. Burning wood and charcoal (made from wood) expanded its use in this environment, which also quickly impacted Europe's supply of wood. During this early period of industry, Western Europe's forests largely disappeared as they served as the raw material for shipbuilding and metallurgy. This shortage led English ironmasters, however, to utilize a new source of energy that would greatly multiply the scale and scope of industrial potential. The English use of coal and more specifically of coke revolutionized the scale and scope of the manufacturing that followed throughout the world.[5]

Although Western Europe had abundant supplies of ordinary coal, it had proven useless for smelting ore because its chemical impurities, such as phosphorus, prohibited its generation of strong iron. For this reason, smelting was fired with charcoal, which was made from wood. Western Europe's lack of wood made it lag behind other regions during these decades. However, around 1709, Darby, mentioned above, discovered that by partly burning coal he could purify it for further use as coke. Darby released this method as a smelting fuel for iron making to the public in 1750. This process proved to be a launching point for the reliance on fossil fuels that would power the Industrial Revolution.

During this era of proto-industry, roughly from 1500 to 1750, there occurred an assortment of great technological developments but no genuine "revolution" of industrial expansion. In an era in which scientific and technological innovations were often frowned upon and when energies and monetary support were focused on exploring the globe, it is relatively remarkable that any developments occurred at all! Simply, European society of the Reformation was not conducive to new technological developments, and the pressure to conform often slowed technological change and kept the implications of energy development fairly confined.

During subsequent eras, a cultural and social fabric took shape that would support, celebrate, and, ultimately, accelerate applications of the machine. Social changes bore a significant impact on later uses of technology, and industry began to move outside of cities. The nation-states that began to develop slowly became somewhat supportive of select technologies. First to find widespread support were technologies and machines that might be used in battle, including designing fortifications, casting cannons, and improving naval fighting ships. But more important to most members of general society, during the eighteenth century, a series of inventions transformed the manufacture of cotton in England and gave rise to a new mode of production that became known as the factory system. Based on a series of related innovations, the new factory-based society that took shape made machines part of nearly every worker's life, tying their work to expanding applications of energy technology.

During these years, other branches of industry stimulated comparable advances and all these together—mutually reinforcing one another—made possible an entire era grown, at least partly, on the back of technological gains. In short, the age would be organized around the substitution of machines for human skill and effort. Heat made from inanimate objects took over for animals and human muscle. Furthermore, this shift enhanced the amount—the scale and scope—of the work that could be undertaken. That was the multiplying effect that became "revolutionary."

After 1750, of course, innovations such as the steam engine generated a bona fide Industrial Revolution. As economic historian Joel Mokyr has written, "If European technology had stopped dead in its tracks—as Islam's had done around 1200, China's by 1450, and Japan's by 1600—a global equilibrium would have settled in that would have left the status quo intact."[6] Instead, in the next two centuries human life changed more than it had in its previous seven thousand years. At the root of this change lay machines and an entrepreneurial society committed to applying new technologies to everyday life, each one relying on new, flexible, and expandable sources of energy.

Particularly informed by these distinctly different outcomes, we can see that the expansion of the factory system based on water power in Britain and the United States established a critical framework for the subsequent industrial expansion that involved fossil fuels. Although the prime mover of the textile industry grew from the technology of an earlier, limited stage of industrialization, businessmen pursued a level of scale and complexity that was new and, indeed, revolutionary.

Starting from 1086 when the Domesday Book accounted for nearly six thousand mills in southern and eastern England, vertical and overshot wheels captured river flows for grinding or cutting. Design innovations such as the undershot wheels allowed early engineers to create a packaged technology that might be placed on any suitable river or stream, and this potentiality guided communities to situate towns accordingly. Smil writes that "more water power capacity was added during the first six decades of the nineteenth century than ever before, and most of these machines continued to operate even as steam power, and later electricity, were conquering the prime mover markets."[7] The expanded scale resulted in textile plants such as those in Lowell, Massachusetts, which became America's first fully integrated clothmaker in 1823. Entirely commodified, the flow of the Merrimack River became the energy source for vast and revolutionary factories that employed thousands. Similarly, by 1840 the MW Shaw waterworks at Greenock, Scotland, ran thirty wheels, and in 1854 the world's largest wheel, Lady Isabella, opened on the Isle of Man to pump water out of the Laxey coal mines.[8]

A NEW VALUE FOR SEA COAL

Similar to others, the transition to energy from coal is riddled with cultural, legal, and social baggage that encumbered its emergence for centuries. Coal was found and mined in many parts of England during the 1200s, with Newcastle emerging as the leading source. While enterprises turned up near the mines, the River Tyne allowed some coal to be dispersed to other locations. While early miners struggled against water to keep the underground mines from flooding, Freese adds that "England's broad use of coal was possible only because the nation had plenty of water to float the coal to market."[9]

The greatest limitation on coal's development in this early era proved to be a basic one: most of the coal along the Tyne lay beneath property held by the Roman Catholic Church. In the Newcastle area, for instance, the church controlled most of the seams that would eventually provide approximately half of Great Britain's entire output. And, simply, the church was reluctant—nearly disinterested—in developing the resource. Certainly, it lacked the drive to infuse the larger cultural and technological innovations that would empower the later revolution. Primarily, miners were serfs—virtual slaves—who divided time between essential farming and optional mining. In England, monarchs often claimed valuable minerals found on any private land; however, coal was grouped with wood or peat and was left to estate holders. Freese writes of early coal: "This humble and possibly manure-smeared fuel was beneath its notice."[10]

Owners of small businesses (primarily, who had begun as serfs) began actively trying to wrest control of the coal trade from the church prior by the late 1200s, with slow success over the next few centuries. Part of the lack of urgency to spark any massive transition grew from social malaise—even disgust—over the dirtiness of coal. Particularly, the problematic smoke it gave off when burned led humans, from the lowliest serf to Queen Eleanor, to refuse to integrate it into their everyday life. Indeed, most domestic fires for heating and cooking remained much as they had been for humans for hundreds of thousands of years![11]

Beginning in 1285, King Edward I established a variety of commissions to study the problem of coal smoke so that it might be mitigated; however, this was not the primary reason for shifting to the new prime mover. Clearly, the transition from wood to coal derived mostly from problems of depleted supply, particularly in England. However, the pressure on the dwindling wood supply received a bit of a respite with the Black Death of the 1300s, which depleted England's population so significantly that forests rebounded to some extent and no energy transition was needed. Additionally, Welsh observers even associated the appearance of the buboes (blisters) that the bubonic plague produced on the human body with "broken fragments of brittle sea coal" that seethed "like a burning cinder."[12] Such descriptions could certainly not have helped the emergent image of coal!

While the coal near Earth's surface had not changed in thousands of years, the human society interacting with it was radically altered during the 1500s. The transitions that would lead to what is known as the Industrial Revolution began with the growing complexity of mining for coal, but it likely wouldn't have occurred without the marital difficulties of King Henry VIII. As the easiest seams of coal had been mined by the mid 1500s, investment was needed to open new areas. The church had little interest in doing so and was still operating mines through short-term leases. This all changed when King Henry VIII ended his marriage to Catherine of Aragon because it had not resulted in the birth of a male heir. The pope's refusal to grant Henry an annulment resulted in England's revolutionary break with the church, one result of which was that between 1536 and 1539 he dissolved the nation's monasteries and confiscated the property—including some of the nation's richest coal mines. Henry's daughter Elizabeth ascended to the throne in 1558, and she played a leading role in guiding the transition to coal.[13]

BRITAIN HOSTS A MONUMENTAL ENERGY TRANSITION

What historians of technology refer to as the great transition is not necessarily the exact emergence of new technologies in the mid-1700s. In order to reach the point of revolutionizing the ways that humans lived their everyday lives, a "great transition" was necessary in intellectual thought that would compound the industrial shift in energy use that was discussed above. Biomass fuels such as wood and charcoal had been in use for centuries, but they could not support an entirely new infrastructural system of machines. Coal, however, emerged as a prime mover for new industries during the 1600s and over the next century would also alter everyday lives.

After England experienced the serious shortages of wood in the 1500s that were discussed above, most of the existing coalfields in England were opened between 1540 and 1640. By 1650, the annual coal output exceeded two million tons. It would rise to ten million tons by the end of the 1700s.[14] In the new energy resource of coal, industrialists found potential power that far exceeded any sources then in use and the complexities of transitioning to coal touched every aspect of British life. Peasants' domestic comfort, most industries, and the strength of the nation's navy were affected by shortages of timber during the 1500s. Laws were passed to limit the taking of wood and commissions enacted to study the shortages. Freese writes: "As the forests continued to shrink, the fuel shortage might have slowed the population growth of the entire nation. . . . But the energy crisis never got that severe for one reason: coal."[15] Although upper-class users initially resisted heating their homes due to dirt and smoke, by the first few decades of the 1600s rich and poor homes in London were heating with coal.

London grew rapidly during the next century, and energy from coal was a primary reason that it became the largest city in Europe by 1750. The symbol of this transition was the technology that had been developed to make life with coal possible: the chimney. Hearth fires, which had been developed to move heating and cooking and fires indoors, required chimneys even when they burned wood. In order to accommodate coal burning, home fireplaces and chimneys were made much narrower to provide the proper draw of air and to send the sooty smoke away and into the atmosphere—or so they hoped!

Exactly what was being burned also mattered a great deal. Particularly for home use, coal was often combined with peat to form a charcoal that would burn much cleaner. "Sea coal" got its name from the fact that it was carried by ship from the northern ports nearest the mines to docks on the lower Thames, but it was bituminous, or soft, coal. When James VI of Scotland succeeded Elizabeth in 1603 to become James I of England, he ordered that all fireplaces in London burn coal, and he ordered the import of hard, clean-burning anthracite from his native Scotland. The real revolution of coal had begun, but accessing it would require a bit more innovation.

CURRENT: Implications of Burning Coal

In hindsight, the historical record notes a few early voices who noted the potential pitfalls of burning coal.

Fumifugium, 1661, John Evelyn

The book *Fumifugium*, an excerpt of which is often included in ecology textbooks today, is treated as the first protest of pollution. In it, English writer and minor government official John Evelyn writes to inform King Charles II of England of his observations. Although his writing elicited little response, he clearly displays a different paradigm for viewing industrial change. Here is an excerpt of his account (with Early Modern English spellings):

IT was one day, as I was Walking
in Your MAJESTIES Palace
at WHITE'HALL (where
I have sometimes the honour to refrefh
my felf with the Sight of Your Illuftrious Prefence, which is the Joy of
Your Peoples hearts) that a prefump-
tuous Smoake iffuing from one or two
Tunnels neer Northumberland'Houfe,
and not far from Scot land-yard y did fo
invade the Court; that all the Rooms,
Galleries, and Places about it were
fill d and infefted with it, and that to
fuch a degree, as Men could hardly
difcern one another for the Clowd,
and none could support, without ma-
nifeft Inconveniency. It was not this
which did firft fuggeft to me what I
had long since conceived againft this
pernicious Accident, upon frequent
obfervation; But it was this alone,
and the trouble that it muft needs pro-
cure to Your Sacred Majefty, as well
as hazzard to Your Health, which
kindled this Indignation of mine a-
gainft it, and was the occalion of what
it has produc'd in thefe Papers.

Your Majefty who is a Lover of
noble Buildings, Gardens, Pidures,
and all Royal Magnificences, muft
needs defire to be freed from this pro-
digious annoyance; and, which is fo
great an Enemy to their Luftre and
Beauty, that where it once enters
there can nothing remain long in its
native Splendor and PerfeSion: Nor
muft I here forget that Illuftrious and
divine PrinceflTe, Your Majefties only
Sifter, the now Dutcbeffe oi Orleans y
who at her Highmfe Jate being in this
City, did in my hearing, complain of
the Effects of this Smoake both in her
Bread and Lungs, whilft She was in
Your Majefties Palace. I cannot but
greatly apprehend, that Your Majefty (who has been fo long accufto-
md to the excellent Aer of other
Countries) may be as much offended
at it in that regard alfo; efpecially
fince the Evil is fo Evidemicall; indan-
gering as well the Health of Your
SubjedSj as it fullies the Glory of this
Your Imperial Seat.

Sir,! prepare in this fhort Difcourfe,
an expedient how this pernicious Nui-
fame may be reformed, and offer at
another alfo by which the Aer may
not only be freed from the prefent Inconveniency; but (that remov'd) to
render not only Your Majefties Palace,
but the whole City likewife, one of
the fweeteft, andmoft delicious Ha-
bitations in the World; and this with
little or no expence; but by improving
thofe Plantations which Your Maje-
fly fb laudably afFeds. . . .
shows that 'tis the sea-coal smoke
That always London does environ,
Which does our lungs and spirits choke,
Our hanging spoil, and rust our iron.
Let none at *Fumifuge* be scoffing
Who heard at Church our Sunday's coughing.[16]

Arrhenius and Krakatua Shape the "Greenhouse Effect" Concept

Momentum had a great deal to do with the Industrial Revolution. It is important to note that early scientists became aware of the implications of pollution from burning coal by the late 1700s. Such insights, though, were largely overwhelmed by the rapid economic growth associated with burning fossil fuels. Experimentation to understand the problem returned at the end of the 1800s.

Building on the early work of French chemist Jean-Baptiste Joseph Fourier and Irish polymath John Tyndall, Swedish chemist Svante August Arrhenius quietly spent his career until his death in 1928 exploring a "hothouse" theory that modeled the impact of pollution on Earth's atmosphere. As scientists began to discuss climate change by 1990, they unearthed one of Arrhenius's papers from the 1890s. Historian of science Gale Christianson writes that, in this paper, "Arrhenius set forth the startling idea that the massive consumption of fossil fuels is capable of raising the temperature of the atmosphere."[17]

When southwestern Indonesia's Krakatau volcano erupted in 1883, a variety of global scientists noted the atmospheric impact of the emissions. Referred to as the Krakataun winds, fine volcanic dust particles circled the planet repeatedly and diffused the Sun's light. They speculated that similar events in the past had contributed to periodic advances of glacial ice. Arrhenius, though, emphasized the contents of the emissions: "The more volatile constituents, such as carbonic acid, sulphuretted hudrogen, and hydrochloric acid, may spread over large areas and destroy all living things by their heat and poison."[18]

During the following decades, Arrhenius devised a theoretical model that used the term *hothouse* to describe what years later would be referred to as the "greenhouse effect." Focusing particularly on the gases of water vapor and CO_2, Arrhenius's calculations demonstrated how significant change in the atmospheric amounts was sufficient to trigger global variations in temperature. The extension of his hypothesis was that an overwhelming change in these gases could spur an event such another ice age. While the technical details appeared largely undeniable, Arrhenius argued in an era that was still very skeptical of humanity's ability to have an impact on the entire globe. In his 1908 publication, he estimated that it would take another three millennia of burning fossil fuels to double Earth's supply of CO_2. For his scientific cry from the wilderness, Arrhenius was awarded the Nobel Prize in chemistry for 1903. Despite such recognition, in making such predictions Arrhenius and other early scientists overlooked the actual, incredible—indeed, revolutionary—expansion of fossil fuel use and the innovations that it would make possible.

EXPANDING COAL'S CAPABILITIES AND IMPACTS

The steam engine possessed a flexibility that allowed it to use coal's energy for a rapidly expanding universe of purposes—thus meriting the term *revolution* for the industrialization that transpired over the next century. In the mining districts, the expansion of labor during the 1700s fueled a new economic and social stratification. Historian Carol Freese writes:

> Coal created a new gulf between classes. The medieval peasants and artisans, whatever their disabilities and trials may have been, were not segregated from their neighbours to anything like the same extent as were the coal miners of the seventeenth century in most colliery districts.[19]

This new laboring class would not discover its ability to organize in order to make workplace demands for decades. In locations such as Scotland, miners were bound into a type of industrial serfdom to coal owners, and their families remained in servitude for generations.

Dismal workplaces, these early mines possessed universal danger of various sorts. It may have been the most dangerous profession of an era in which there were many. Miners worked in utter darkness in fear of fire, drowning, ceiling collapse, and a lack of oxygen, just to name a few of the challenges atop grueling labor in cramped conditions. Referred to as "choke damp," carbon dioxide (and other lethal gasses) or the "white damp," which was carbon monoxide, could build up in the mine and poison workers with a startling immediacy or a deceptive gradualness. It was the insidious white damp that prompted miners to bring a small caged animal (such as a canary or mouse) into the work site to provide a warning of poisonous gas. And *fire damp* was the term for the hissing seeps of flammable methane that would leak from the coal seams and catch fire or explode in contact with lantern flames. Of course, we need to recall that the only workplace lighting in mines came from open flames. In regions such as Newcastle, catastrophic mine explosions were commonplace.

Beginning from the use of pumps and engines, flooding dictated how rapidly the industry expanded through the 1700s. The expense of such novel machines accelerated the trend toward even larger mines and workforces, which was the primary way that companies could recover the large outlay that such equipment required. This perpetual cycle of expansion and growth rapidly made mining Britain's largest industrial enterprise. By 1700, coal production in Britain had grown ten-fold from its standing in 1550 and the potential power of coal far exceeded what wood could have provided. It is estimated that by 1700, Britain mined five times more coal than the rest of the world combined.[20] The search to properly connect the power of coal to profitable use became the defining characteristic of the early Industrial Revolution.

The coal, though, that their labor produced was transforming the world and providing raw material to define what it meant to be among the world's "haves." The interconnection of society's use of coal led the author Daniel Defoe to write of Newcastle:

> Whereas when we are in London and see the prodigious fleets of ships which come constantly in with coals for this increasing city, we are apt to wonder whence they come and that they do not bring the whole country away; so on the contrary, when in this country we see the prodigious heaps, I may say mountains of coals which are dug up at every pitt, and how many of these pitts there are, we are filled with equal wonder to consider where the people live that can consume them.[21]

The ability to make steam by burning the coal proved to be the great multiplier to power this energy transition. James Watt and others took the basic steam engine design and began applying it to diversified industrial outputs. Still with coal heating its steam, these engines squeezed more than four times as much motive force from it by the 1770s. Applied to purposes such as manufacturing iron, pumping water, and, eventually, to transportation, Watt's engine allowed the flexibility that spread across societies. Steam increased the demand for both coal and iron while also making each item easier to produce. The multiplying effect of making coal cheaper attracted more people and purposes to steam power and became the force behind a true revolution in how humans lived. Freese writes: "Steam power did not create the factory system, but it irreversibly changed the scale, the nature, and the location of industrial enterprise."[22]

In Britain, industrialization had taken root without coal power, particularly in the water-powered textile factories (discussed in chapter 2). Innovators such as Richard Arkwright and Samuel Crompton fashioned complex systems of looms and spinning spools powered only by simple waterwheels. The system of jennies grew and the mills employed thousands; however, the steam engine also soon raised this industrialization to new levels. In Britain, for instance, the potential of steam exasperated the limitations of water-powered industries such as milling. No longer did factories need to be located only along riverways where dams and other infrastructure could be constructed.

In London and its environs, the symbols of the new age of coal power were gargantuan applications of the same chimney accommodations made to British homes. By the mid-1800s, these industrial chimneys were springing up at a rate of more than a hundred per year. Transforming the cityscape, the stacks reached well over three hundred feet. Made of brick, such a chimney required five hundred thousand to a million of them and was held together by a slow-drying, lime-based mortar. They needed to withstand gases as hot as 1,200 degrees F. The taller the chimney, it was thought, the more likely it was to allow the dirty smoke to exit the immediate area and move off into the atmosphere. While this was not based in a full understanding of industrial pollution, these chimneys suggest a very early awareness that the new, coal-powered age of industry had its drawbacks.

By the mid-1800s, the size of a community's industrial chimney had become a badge of honor. For instance, in 1857 the Port Dundas Chemical Works in Glasgow erected a chimney that rose 454 feet (for reference, this was approximately thirty feet less than the Great Pyramid of Khufu). To reach this height, they built a solid limestone foundation on thirteen acres. Although the chimney

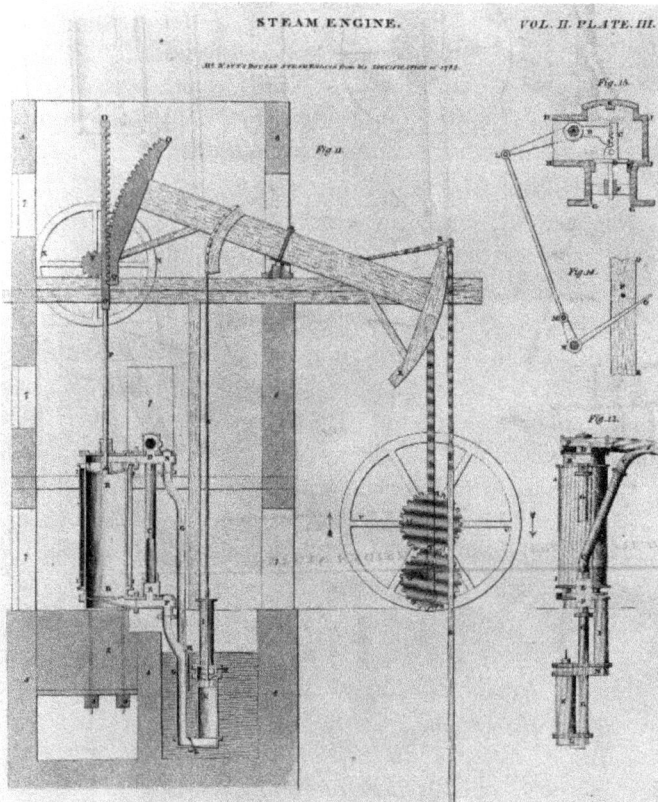

Steam engines were a technology that relied on coal for fuel but also enhanced humans' ability to increase supply by draining flood waters from underground mines. LIBRARY OF CONGRESS PRINTS AND PHOTOGRAPHS DIVISION

had an outside diameter of thirty-two feet at its base, it tapered as it rose until the diameter was only thirteen feet at the crown. Such structures also carried great risk and often collapsed. Known as "a fall," chimney collapses often occurred during or immediately after construction due to the lack of secure ground beneath its foundation. The same wind that was meant to carry away the poisonous effluent, however, also blew over many chimneys. This danger led many engineers to emphasize round chimneys over square ones.[23]

"The most prominent feature of the world at large," wrote one engineer in 1889, "of every stream or power plant, and that by which the manufacturing character of a village or city is most easily distinguished, is the chimney."[24] As the essential backdrop to the industrial factory system, the expansion continued when the American firm Weber Company erected over a thousand cement chimneys between 1903 and 1910. Gone were the symbols of the Victorian era made of brick and mortar, replaced by metal and cement monsters.

The Steam Engine Enables Industrialization

In order to increase coal supplies to a level that could revolutionize human life, mine owners needed to tame the mines through which humans acquired the mineral. In this unique case, one device—the steam engine—helped to achieve this while also serving a host of other applications that powered industrialization.

The basic idea of the steam engine had grown from some of the revolutionary intellects of this new era who were becoming more and more free to openly explore innovations that might significantly alter the very fabric of human life. For instance, the idea of the piston, which was the basis of the engine, only came about after the realization of the existence of Earth's atmosphere. Although other societies had thought about the concept of an atmosphere and pressure holding things to Earth, it was Europeans who began to contemplate the possibilities of replicating this effect in miniature.[25] During the mid-1600s, English engineers experimented with a machine that utilized condensation in order to create a repeating vacuum to yield an output of motive power. The first model of such a device is attributed to Denis Papin, who in 1691 created a prototype piston that was moved within a cylinder by using steam. This device remained unreliable for commercial use, though, because the temperature could not be controlled. The piston, and therefore the engine, could only be applied if it was continuous and reliable and, therefore, produced consistent force.

In 1712, Thomas Newcomen used atmospheric pressure in a machine that he alternatively heated and cooled in order to create the condensation pressure necessary to generate force. An ironmonger from Dartmouth, Newcomen's family had lost its property in the wake of Henry VIII's revolutionary changes. The engine that he perfected had a specific purpose in his family's holdings: to remove the continuous seep of water that inundated their coal mines, flooding them and making them useless and their coal inaccessible. His successful engine was used exclusively for removing the water from inundated coal mines. Primarily, Newcomen had accidentally discovered that a small jet of cold water introduced into the single piston as it reached a full head of steam created the pressure to push the piston down, which in turn raised the suction pump rod and lifted ground water to the surface. Placed directly above the mine shaft and staffed by men who fed coal into the boiler, the steam engine achieved its goal. As a machine, the early steam engine was impossibly inefficient and its need for constant fueling meant that a coal mine was the *only* suitable site for its use.

Newcomen's engine was fairly simple to replicate by English craftsmen and it quickly spread to Belgium, France, Germany, Spain, Hungary, and Sweden by 1730. Although it lacked efficiency and could not generate large-scale power, the Newcomen engine was a vision of the future. It marked the first economically viable machine to transfer thermal energy into kinetic energy. This concept, powered by a variety of energy sources, was the flexible prime mover that would lead the Industrial Revolution.

In the short term, though, the steam engine allowed access to vast new stores of anthracite coal. Throughout the mining region, the problem of underground seepage plagued miners and mine owners, particularly in those that had been sunk along the coastline. The steam engine solved this problem, allowing the new development of abandoned mines and massively expanding the overall supply of coal.

ANTHRACITE COAL MINING AND AMERICAN EXPANSION

Into this expanding capacity went a massive supply of coal. Particularly in the United States, anthracite, or hard coal, proved most transformational. However, the process was rooted in the basic European settlement starting in the 1600s. Linked by ships, European powers sought necessary resources in other regions. Soon, this led the mercantilist nations to establish colonies. In North America, settlement grew from agriculture; however, as the United States developed, it emphasized industries—using technologies perfected in Europe and new ones that blazed important new paths. The key connecting each undertaking was that energy was the necessary raw material to develop the young nation. Known as the "grand experiment," the United States functioned almost as a laboratory for expansively applying industrial ways to developing an entire nation.

America of the early 1800s still relied on energy technologies that would be considered sustainable alternatives to fossil fuels. The transition, though, had begun as industrialists expanded the use of charcoal, which created an infrastructure that could be expanded to include additional energy sources. Some of these resources, however, were complicated to harvest and manage. Their acquisition demanded entirely separate technological innovations as well as shifts in the accepted patterns of human life. Surface coal in North America, wrote one theologian, had been "scattered by the hand of the Creator with very judicious care, as precious seed, which though buried long, was destined to spring up at last, and bring forth a glorious harvest."[26]

In the early 1800s timber or charcoal (made from wood) filled most Americans' heating and energy production needs. This changed rather suddenly during the War of 1812, which pitted the United States against Great Britain in a conflict over trade and ended in stalemate in 1815. Discussed above, the conflict was rooted in American sailors being impressed to serve in the British Navy. On the sea, though, Britain's major maritime initiative was the disruption of trade into American ports. In Philadelphia, for instance, the British blockade nearly crumbled the economy of the young republic, which was discussed in chapter 2.

As the Industrial Revolution swept from Europe and into other parts of the world in the early to mid-1800s, the nations most susceptible to its influence were rich in raw materials and committed to the individual freedom of economic development. In these considerations, the United States led the world. Thanks to the American interest in free enterprise and the astounding supplies of raw materials, including coal and later of petroleum, the United States became the industrial leader of the world by the early 1900s—after only four or five decades fully committed to industrialization. Economic prosperity, massive fortunes for a few, and employment for nearly everyone who wanted to work were a few of the outcomes of American industry. Another outcome, though, from the intense use of the natural resources exerted by industrialization was environmental degradation.

In the industrial era that stretched from 1850 to 1960, many American industrialists were willing to create long-term environmental problems and messes in the interest of short-term gain. Some of these gains came in the form of unparalleled personal fortunes. Other benefits included long-standing economic development for communities and regions around the United States. However, this economic strategy took shape on the back of the harvest, manipulation, and exploitation of natural resources. This ethic of extraction was felt to some degree in any industrial community, but possibly it was most pronounced in mining areas, particularly those areas mining for energy resources such as coal and petroleum.[27]

As American society committed to a primary course of development that was powered by fossil fuels, much of the evidence of extraction and production was viewed as symbols of progress. Few checks and balances existed to demand care and conservation. In the nineteenth century, the environmental consequences of mining for these hydrocarbon resources buried deep in Earth was of little concern. Most often, industries were viewed almost solely for the economic development that they made possible. Smoke, pollution, and the effluent of industrial undertaking marked progressive symbols of economic expansion.

Railroads Link Enterprises for Expansion

In addition to stimulating the development of mining in locales such as Pennsylvania, industrial development contributed to and fed the development of related undertakings. More and more industries became essential to everyday American lives. In most cases, each of these undertakings derived from burning coal or other fossil fuels, therefore broadening the impact of emissions. Throughout American history, transportation was one of the most important applications of energy use. In the case of coal, the use of the railroad made coal supplies accessible while also involving coal's energy in innumerable other activities during the 1800s.

The planning and construction of railroads in the United States progressed rapidly during the nineteenth century. Some historians take the view that this occurred too rapidly. With little direction and supervision from the state governments that were granting charters for construction, railroad companies constructed lines where they were able to take possession of land or on ground that required the least amount of alteration. The first step to any such development was to complete a survey of possible passages.

Before 1840, most surveys were made for short passenger lines that proved to be financially unprofitable. Under stiff competition from canal companies, many lines were begun only to be abandoned when they were partially completed. The first real success came when the Boston and Lowell Railroad diverted traffic from the Middlesex Canal in the 1830s. After the first few successful companies demonstrated the economic feasibility of transporting commodities via rail, others followed throughout the northeastern United States.

The process of constructing railroads began with reconstruing humans' view of the landscape. Issues such as grade, elevation, and passages between mountains became part of a new way of mapping the country. Typically, early railroad surveys and their subsequent construction were financed by private investors. When shorter lines proved successful, investors began talking about grander schemes. These expansive applications of the railroad provided the infrastructure for remarkable commercial growth in the United States, expanding the impact of the Industrial Revolution.[28]

By the 1850s, though, the most glaring example of this change was coal-powered railroads. The expanding network of rails allowed the nation to expand commercially. Most important, coal-powered railroads knitted together the sprawling country into a cohesive social and commercial network. Although this could be seen in microscopic examples, including cities such as Pittsburgh and Chicago to which railroads brought together the raw materials for industrial processes such as steel making, on the macroscopic scale railroads allowed American settlement to extend into the western territories.[29]

It was a cruel irony that the industrial era that evolved in the late 1800s relied intrinsically on transportation. Long, slender mountains stretched diagonally across Appalachian regions such as Pennsylvania, creating an extremely inhospitable terrain for transporting raw materials. Opening up the isolated and mountainous region required the efforts of a generation of capitalists and politicians, who used their resources and influence to create a transportation network that made the coal revolution possible. Canals were the first step in unlocking the great potential of the coalfields. Soon, though, industrialists focused on a more flexible transportation system that could be placed almost anywhere. Railroads quickly became the infrastructure of the industrial era. Knitting together the raw materials for making iron, steel, and other commodities, railroads were both the process and product of industrialization.[30]

The iron rails produced in anthracite-fueled furnaces extended transportation routes throughout the nation. This revolution in transportation led to corresponding revolutions in the fueling of industries and the heating of urban residences, which in turn required more and more miners and laborers. Although each of these social and cultural impacts of the railroad altered American life, it was, after all, primarily an economic enterprise. Primitive as it was, the antebellum railroad entirely remade American commerce, particularly reconfiguring the very ideas of prices and costs. Previously, prices had factored in the length of time involved in transporting goods via turnpikes, the steamboat, and the canal. From the start, railroad rates were significantly cheaper than wagon rates. The increasing systemization of the railroad process made low costs even more possible.[31]

The possibility of railroads connecting the Atlantic and Pacific coasts was soon discussed in the Congress and this initiated federal efforts to map and survey the western United States. A series of surveys showed that a railroad could follow any one of a number of different routes. The least expensive, though, appeared to be the 32nd parallel route. The Southern Pacific Railroad was subsequently built along this parallel. Of course this decision was highly political; southern routes were objectionable to northern politicians, and the northern routes were objectionable to the southern politicians.

Although the issue remained politically charged, the Railroad Act of 1862 put the support of the federal government behind the transcontinental railroad and clearly linked American economic development to coal. This act helped to create the Union Pacific Railroad, which subsequently joined with the

Central Pacific at Promontory, Utah, on May 10, 1869, and signaled the linking of the continent.

Railroading became a dominant force in American life in the late nineteenth century, and one of its strongest images was its ability to remake the landscape of the entire country. Following 1880, the railroad industry reshaped the American built environment and reoriented American thinking away from a horse-drawn past and toward a future with the iron horse.

Steel Manufacture Provides Infrastructure

Railroads and the reliance on burning coal for power enabled the implementation of complex industrial undertakings at a scope and scale never seen before. Although iron manufacture increased in scale with the more intense model of industrialization after 1850, steel is possibly the best example of this new era's capabilities. Using railroads as its linking device, Andrew Carnegie perfected the process of steel manufacturing and created one of the greatest fortunes in history.[32]

Into one pound of steel, observed Carnegie, went two pounds of iron ore brought 1,000 miles from Minnesota, 1.3 pounds of coal shipped 50 miles to Pittsburg, and one-third of a pound of limestone brought 150 miles from Pittsburgh. Rivers and railroads brought the material to the Carnegie Steel Works along Pittsburgh's Monongahela River where Bessemer blast furnaces fused the materials into steel. One of the greatest reasons for the rapid rise of American industry was its flexibility compared to that of other nations. Railroading could be integrated immediately into various industries in the United States, which, for instance, allowed American industry to immediately embrace the new Bessemer steelmaking technology. Other nations, such as Britain, needed to shift from previous methods.

One innovation contributed to another in the late industrial era. Inexpensive energy made it feasible to gather the disparate materials that were necessary to make steel. Steel was stronger and more malleable than iron, which made possible new forms of building. Carbon levels make the bulk of the distinction between the two metals. Experiments with removing the oxygen content of pig iron required more heat than ordinary furnaces could muster. The Bessemer invention created a "Bessemer blow," which included a violent explosion to separate off additional carbon and produce the 0.4 percent oxygen level that was desirable for steel.

New tasks, such as running the Bessemer furnace, created specialized but also very dangerous jobs. Working in the steel mill created a new hierarchy for factory towns. In the case of steelmaking, hot or dangerous jobs such working around the Bessemer furnace often eventually fell to African American workers after the era of slavery.[33]

CURRENT: Dynamos Create New Flows of Energy

In the United States, industrial development took on the level of a new religion. For instance, starting in May 1876, nine million observers streamed to the Centennial Exhibition in Philadelphia. At the opening ceremony, a choir sang Handel's "Hallelujah Chorus" and U.S. president Ulysses S. Grant pulled the giant control lever that started the mammoth Corliss steam engine and huge hiss of released pressure followed. Throughout the great hall, visitors saw the diverse progress embodied by countless mechanical innovations—each one spilling, sawing, sewing, pumping, or printing with barely the need for human workers. Instead, the motive force of the single great coal-powered engine drove each machine and catapulted humanity forward in a bold, new fashion.[34]

A precursor of the dynamos that would follow, the Corliss engine set the parameters for the next great leap that would revolutionize human work: flexible forms of power that could be applied to endless tasks. In particular, electricity—an artificially generated power source—made steam generated from burning coal, then run through a dynamo that converted the steam's heat into electric current into an entirely separate and eminently expandable source of power. Over the next decades, lighting and streetcars were electricity's primary applications due to the fact that conversion required costly infrastructure. In these cases, public partnerships combined with corporate initiatives to help make the transition easier to afford.

By 1900, factories rapidly began to electrify their operations. Environmental historian Chris Jones writes: "In 1899, fewer than 5 percent of industrial motors ran on electricity steam engines and waterwheels were much more common. By . . . 1914, electricity supplied nearly two-fifths of the nation's industrial power." The war effort stimulated conversion, therefore, and by 1919 "electricity was the nation's primary source of industrial power."[35] Compared to other sources of power, electricity was becoming much more plentiful and expandable and factory owners viewed it as safer, quieter, more reliable, and more efficient.

For industrial growth in the United States and the world, Jones explains, electricity made power tangible and manageable. Most consequential, he writes, "was that a single block of power could be subdivided and easily transmitted to multiple work stations."[36] With previous forms of energy, the transfer of power—distributing it from the central waterwheel or steam engine—was a separate and expensive engineering feat. From the metal shafts, leather belts and pulleys, and levers of the past, cables and wire (either hung above or buried in the ground) now literally and figuratively hid the power from view. It could be distributed behind the scenes—almost by

In certain examples, such as the 1876 Corliss steam engine, technologies became heroic symbols of the new industrial age. LIBRARY OF CONGRESS PRINTS AND PHOTOGRAPHS DIVISION

magic—and appear elsewhere. And yet it still originated in a hunk of coal. Finally, electric motors allowed industrialists to manage or control the rate of power flowing to each task. Although we take this variability for granted, it proved absolutely revolutionary to humans' life with energy.

Electricity, therefore, was not just another source of power—its unique properties transformed manufacturing. Tracing its power backward, electricity also achieved this transformation in the name of fossil fuel burning. In coming decades, high-voltage transmission methods and wiring spread these capabilities throughout the world and allowed electricity to take on the ability to alter living patterns so much so that it became one of the symbols of modern society, suggesting which nations and communities were progressing and which ones were not.

CONCLUSION: MINING EXPANDS SCALE AND SCOPE WORLDWIDE

While the United States remained the global leader in coal production through 1950, iron ore and therefore the production of metals shifted to the Soviet Union by the 1930s and then by the end of the twentieth century to China, Brazil, Australia, and Russia.[37] From early efforts using pick and shovels, mining expanded through new tools and tactics to include strip and opencast mining, which focused on beheading mountains in order to access the minerals. On the ground, the additional environmental impact proved profound. The emphasis on burning the massive amounts of cheap coal unearthed by these methods over the course of the 1900s also expanded the scale of the local pollution issues that had been evident as far back as Queen Elizabeth in the 1600s.

McNeill writes that by 1870 "Britain had perhaps 100,000 coal-fed steam engines, all churning out smoke and Sulphur dioxide. The English Midlands became known as the Black Country," and an estimated quarter of the deaths came from lung diseases.[38] From this start, coal-fired industrialization over the course of the era of high energy use expanded through Europe, the United States, and Japan and became a clear demarcation of the "haves" on the global scale. Cities in North America and Europe acquired electric grids powered by coal combustion in the 1910s and 1920s. More intense periods of industrialization followed: in Eastern Europe after 1940; Japan after 1950; after 1960 in East Asia, particularly South Korea; and in China after 1978. Pollution was a common output, most notably infecting the most ubiquitous sink: Earth's atmosphere. Restraints on air pollution mattered little in any of these cases prior to 1980.

While the revolution known as industrialization had begun in Europe and elsewhere, the profound scale of enterprise had exploded through the American method of work and production. The revolution had been in how products could be produced, and fossilized energy—particularly coal—proved to be the separation between the biological old regime and something different. The next chapter emphasizes how this high-energy capacity moved from the sector of production to also impact the everyday existence of humans in nations possessing access to abundant energy.

NOTES

1. Barbara Freese, *Coal* (New York: Basic Books 2016), p. 19.

2. Freese, pp. 20–21.

3. Freese, p. 22.

4. For a general contextual interpretation of this transition, readers should consult Simon Pirani, *Burning Up* (New York: Pluto Books, 2018).

5. See, for instance, Astrid Kander, Paolo Malanima, and Paul Warde, *Power to the People: Energy in Europe over the Last Five Centuries* (Princeton, NJ: Princeton University Press, 2014).

6. Joel Mokyr, *The British Industrial Revolution* (London: Routledge 2018), pp. 52–53.

7. Vaclav Smil, *Energy and Civilization: A History* (Cambridge: MIT Press, 2017), p. 155.

8. Kander et al., pp. 156–160.

9. Freese, p. 22.

10. Freese, p. 23.

11. Kander et al., pp. 60–62.

12. Freese, p. 27.

13. Freese, p. 31.

14. Freese, p. 32.

15. Freese, p. 32.

16. John Evelyn, *Fumifugium* (United Kingdom: University of Exeter, 1976). https://archive.org/details/fumifugium00eveluoft/page/n5/mode/2up. Accessed on January 6, 2022.

17. Gale Christianson, *Greenhouse* (New York: Walker and Co., 1999), p. 106.

18. Ibid., p. 113.

19. Freese, p. 45.

20. Freese, p. 56.

21. Freese, p. 44.

22. Freese, p. 67.

23. Christianson, pp. 60–62.

24. Christianson, p. 60.

25. Christianson, p. 60.

26. Freese, pp. 105–106.

27. Brian Black, *Petrolia* (Baltimore: Johns Hopkins University Press, 2000).

28. John Stilgoe, *Train Time* (New Haven, CT: Yale University Press, 2007), pp. 3–8.

29. See John Stilgoe, *Common Landscape of America* (New Haven, CT: Yale University Press, 1983).

30. See William Cronon, *Nature's Metropolis* (New York: Norton, 1991).

31. Cronan.

32. See John Opie, *Nature's Nation* (New York: Cengage Learning, 1998).

33. Opie.

34. Freese, p. 129.

35. Christopher Jones, *Routes of Power* (Cambridge: Harvard University Press, 2016), p. 197.

36. Jones, p. 198.

37. John R. McNeill, *Something New Under the Sun: An Environmental History of the Twentieth-Century World* (New York: Norton, 2001), pp. 32–34.

38. McNeill, p. 58.

4

Energizing Everyday Human Life

GROUND: LUCIFER MATCH

In American Lucifers, historian Jeremy Zallen explains the unique circumstance of fifteen-year-old Richard Toye, who with many other young boy and girl laborers in Manchester, England, in the mid-1800s "glowed in the dark."[1] *These young workers were part of the "violent ecology of phosphorus" that guides us modern readers to begin to grasp what it required to bring ubiquitous light to a dark world. Toye and others were part of the industrial process to create matches that cost so little that they changed humans' relationship with fire and particularly its by-product, illumination. Revolutionary in its own right, access to ubiquitous light also moves nineteenth-century energy relations beyond the industrial workplace to an all-important new connection to the domestic consumer.*

Phosphorus began its transformation to light throughout the world wherever bone or bird waste—guano—could be collected in massive amounts. In Lyon, France, and Birmingham, England, manufacturers transitioned from bone from various animals to phosphate-rich rock guano mined on West Indian islands such as Sombrero.[2] *The poison material held great danger for workers such as Toye, but the era saw workers largely as cogs in the industrial machine. "No nineteenth-century job," wrote one historian, "was as difficult, dangerous or demeaning as shoveling either feces or phosphates on guano islands."*[3]

Regardless of its form of origination, the phosphorus arrived at match factories in England, France, and later the United States, many of which could reach a capacity of producing 500 splints or matches per second, 3.6 million per hour, and more than 10 billion in a year by the mid-1800s. One of the first manufactured products that could be produced on a mass scale, these matches had the capacity to alter human existence. Each match, often referred to as a "lucifer," of course, made a flame readily and repeatedly available to its user—whether it would be used to light a cigar, a stove, or a lamp. "Lucifer matches may have been small, cheap things, but they remade consumers' worlds, and people knew it," writes Zallen.[4]

In reality, the lucifer matches involved each consumer in a revolutionary energy exchange. In fact, they became the holder of vast energy potential. As consumers, the choices brought directly to them spread broadly from the ignition of a match. Their new view of modernity expected raw power to be available for everyday activities, whether for personal convenience or efficiency. Energy was no longer reserved only for industrial endeavors, and consumers clamored for new, labor-saving conveniences!

Unlike any structure before it, the Crystal Palace that opened on May 1, 1851, as the centerpiece of the Great Exhibition in London remained standing only through the combined miracles of human ingenuity in the industrial era. From its intricate network of slender iron rods that held upright massive sheets of clear glass, the Crystal Palace's main building measured 1,848 feet long and 408 feet wide and rose to 108 feet in its central transept while occupying some eighteen acres. On the ground floor and galleries there were more than eight miles of display tables containing gadgets and innovations from around the world.

In retrospect, forming a collective, global vision of progress proves to be as important to expanding the high-energy existence as any specific new innovations. Whether other nations specifically sought to construct their own crystal palaces, the symbol of this structure outside London and the event of a global exposition helped to establish a vision of the future that guided the large-scale adoption of new ways of living. It is this spirit—and the assumption of energy supplies at its core—that can be found in the image of structures, designs, and ideas that became emblems of modernization and representations of whether nations fit into categories of "haves" or "have-nots."

The British construction effort of the Crystal Palace began in 1849 when Prince Albert, husband of Queen Victoria and president of the Royal Society of Arts, conceived the idea of inviting international exhibitors to participate in an exposition. Designed by Sir Joseph Paxton, the Crystal Palace held approximately fourteen thousand exhibitors (nearly half of whom were non-British), including manufacturers of false teeth, artificial legs, Colt's repeating pistol, Goodyear India rubber goods, chewing tobacco, McCormick's reaper, hydraulic presses, powerful steam engines, pumps, automated cotton mules (spinning machines), and wooden or paper matches. More than six million visitors attended the exhibition, gawked with wonder, and then returned to their homes with new expectations and ideas about their future. The Crystal Palace established an architectural standard for later international fairs and exhibitions that likewise were housed in glass conservatories. Just like the Great Exhibition of 1851, these events became a tool of modernization—a kind of demonstration zone for new ideas and innovations.

As access to energy expanded and powered new factories and changes in the workplace, an industrial revolution transformed economies and allowed societies to differentiate their productive capabilities. The Great Reversal in world economies grew from these beginnings in Britain and elsewhere before becoming a platform for international expansion in nations that pursued industrial development. Industrialization, then, allowed a few to accrue significant wealth; eventually, humans' overall standard of living also improved through laborers' access to stable incomes. In many of these societies, such a new economy led to related revolutionary changes in basic patterns of human living.

Sometimes, the connection of new technologies with everyday life came in overt moments such as the Crystal Palace in which developing nations sought to display their innovations to their own citizens and the world. In other examples, though, innovations were presented at a more grassroots level to consumers through developing markets designed for individual consumption. By the end of the nineteenth century, the great technological innovations of specific inventors became cogs in broad-based systems that could be applied to transforming living patterns of an entire nation—even for all of humanity. Historian of technology Thomas P. Hughes explains that such system builders focused on a nation such as the United States where they

> found that a nation committed to mass consumption, freedom of enterprise, and capitalism particularly suited their goal of technological-system building, whether it was socially benign or destructive. Some were motivated by desire for power and money, but they shared a drive to order, centralize, control, and expand the technological system over which they presided.[5]

On the global scene, the United States at the end of the nineteenth century presented the perfect laboratory for such modern ideas where the flexible markets of capitalism directly connected innovators to mass consumers. A dominant theme of the twentieth century, then, becomes the way that other nations draw from these efforts to centrally modernize their own societies.

Clearly, the late 1800s began an era—extending through the twentieth century—when national control and government could be applied to stimulate development that was seen to be in the long-term interest of a nation. Power, particularly in the form of electricity, was at the center of this critical era of modernization when new technologies moved beyond factories and into everyday lives. Creating personal sources and applications for energy made standards of living in some Western nations radically different from that of humans living elsewhere. To exemplify this, this chapter considers a variety of examples of technologies that radically modernized human life; however, the element around which it is framed—starting from the widely available match—is illumination.

Tracing the radical alteration of lighting between 1850 and 1930 demonstrates the pace of technical innovation and also the formation of a massive gap between societies with and without access to illumination technologies.

In general, though, this chapter focuses on the basic trend from the late nineteenth century forward of democratizing a high-energy existence for humans that grew directly from industrialization and differed significantly from earlier living patterns. On the macro level, new sources of energy were applied in transportation sectors to provide a strategic advantage for national growth, while at the micro level similar innovations altered individuals' expectations for basic portions of life, including food, personal safety, and individual concepts of space and time. In short, a vast portion of the human population adopted radically new ideas about our species' capabilities after 1850. By adopting these new patterns, humans created a significant gap separating haves from have-nots. And they solidified new basic ideas of time, space, movement, and health that redefined expectations for humans in the twentieth century. Such advancements share a common point of departure contained in the Crystal Palace of 1851: a clear identification and acceptance that energy was essential to human life, and the more the better.

TRANSPORTATION AS ENGINE FOR GLOBALIZATION

New, flexible infrastructure such as railroads provided an organizational structure to settlement and economic expansion in the nineteenth century. In such a manner, a coal-powered technology that was supported or carried out by a national government could form its system and strategy for growth. Whereas railroads began at a mine in Durham in the north of England in 1825 and expanded to twenty-three thousand miles in England by 1850, their transformational capabilities were fully realized elsewhere. Due to timing and population dispersal, the United States presents the clearest example for how the railroad system provided logic, organization, and stimulation to determining a nation's future. Knitting together the raw materials for making iron, steel, and other commodities, railroads were both the process and product of industrialization. The iron rails produced in anthracite-fueled furnaces extended transportation routes throughout the nation. This revolution in transportation led to corresponding revolutions in the fueling of industries and the heating of urban residences, which in turn required more and more miners and laborers.

Drawing power from prime movers such as coal or wood fed into a steam engine, the impact of railroads in nineteenth-century America functioned on a variety of levels. Initially, small-scale use of railroads allowed specific industries, such as iron and steel, to expand by more easily bringing disparate raw materials to a single site of manufacturing. For instance, Pittsburgh, Pennsylvania, became the global leader for manufacturing these metals in just this fashion. In the American West, though, railroad technology reached new levels of importance by binding together distant regions and resources to enable a new, expanded economy that really could not exist prior to the innovation.

Communication was essential to the railroad's success and a tool for its expansion. Railroads shrunk ideas of time and space by making it possible to trade materials from distant areas relatively quickly. The expanding rail network, however, also possessed its own infrastructure: the telegraph, which was invented by Samuel F. B. Morse in 1844. From the fateful day when Congress rejected the idea that the telegraph was a logical extension of the postal service, the growth of the new technology was sporadic. How could such a system be financed and made profitable? In the mid-1850s, Hiram Sibley convinced investors that a whole telegraphy system would be more valuable than the sum of its parts. He and his associates built Western Union, which over the next ten years became the hub for the entire American telegraphic system. Focusing on the corridor between Buffalo and Chicago, Western Union became the dominant player in the settlement of the American West. Sibley and Western Union were hired by the federal government to string a wire across the vast American West just prior to the Civil War. As a defining technology of the new railroad era, the telegraph business and Western Union soon became a most attractive investment. The famous American financier Jay Gould soon made Sibley's company one of his most valuable properties.

Survey crews and engineers established the most prescient routes through the American landscape, so it made sense that telegraph lines should follow. Very quickly, though, the telegraph made itself indispensable to the railroad system. By the 1870s telegraphic dispatching had organized the railways into a regimented endeavor on which Americans could depend and around which commerce could structure itself—including the use of train stations as telegraph offices. The telegraph and railroad became a marriage of related technologies. As they passed endless hours crossing the nation, train crews were required to keep a sharp eye out for breaks in either the rail or telegraph system.

Running in synchronicity, the wires and poles extended across the nation, typically taking advantage of the railroad's right-of-way and running directly next to the tracks. This complement to the rail technology shrank time and space further by facilitating communication between stations. This communication, which was considered to be essential for the scheduling and safety of rail travel, also allowed Americans their first long-distance communication network. In 1865, after thirty-five years of steady, cautious development in the Northeast and Midwest, the American railroad system was far from complete. From thirty-five thousand miles in 1865, the network grew to embrace nearly two hundred thousand miles by 1897. As a unit, the trunk line railroads of the Northeast may rank as the most impressive concentrations of economic capital in human history. This infrastructure, then, allowed late nineteenth-century American society to achieve unrivaled industrial and commercial development.

Based on the success of American use of the railroad to power economic expansion, other nations, including Germany, France, and Russia, also integrally used the railroad to stimulate economic expansion by the late 1800s. When Russia possessed just seven hundred miles of railroads in 1860, it used its federal Ministry of Finance to stimulate rapid expansion. By 1900, Russia's rail system had grown to thirty-six thousand miles of rail that could tie far-off areas of Siberia with its industrial hubs.[6]

Annihilating Space in the United States: Texas and Cattle

The main rail line to the West offered the access that was essential to further development, and it almost magically possessed the ability to take ordinary resources in one region and connect them to markets elsewhere in which they possessed greater value and importance. One of the greatest examples of the railroad's magic might be found in Texas, where the main thoroughfare was a grassy strip that led to the north.

Similar to filling stations along the interstate highway, the swath of prairie in the American interior created a transportation corridor for cattle. In this humid grassland, alkaline soils prohibited extensive tree growth. Big and little bluestem, wire grass, Texas winter grass, and buffalo grass rose toward the wide-open sky while the grass's real life—its root structure—extended as much as seven feet into Earth. The expanse of the prairie impressed everyone who saw it; however, only a few observers appreciated it as the raw energy for Texas's future as a cattle empire.

While the railroad was the most important factor in the settlement of the western United States, not every region immediately received rail stops. Cattle proved portable enough to close this divide. However, long trail rides meant that the cattle would lose weight and be worthless by the time they arrived at cattle yards in Omaha and Chicago. Cattlemen searched for grassy corridors that would allow the cattle to munch as they walked to market hundreds of miles away. Throughout the world, humans have long raised cattle for meat. Americans, however, tied cattle raising into larger economic markets. This required transportation; without this conduit, the connection to markets did not exist. Cattle were moved by trail in the eastern United States as early as the late 1700s. The scale of trailing in Texas and other western states expanded immensely by the 1860s. Plus, Texas ranchers moved wilder beasts—longhorn cattle and mavericks. After the Civil War, the longhorns were Texas's primary asset (estimated to number from three to six million, which was more than six times the human population in the state).

Nations organized entire transportation systems around rail travel powered by coal.
LIBRARY OF CONGRESS PRINTS AND PHOTOGRAPHS DIVISION

With access to outside markets cut during the Civil War, postwar trailing in Texas might have ended completely. But Joseph G. McCoy of Illinois actively sought a new model for getting Texas beef to consumers. In the spring of 1867, he convinced officials of the Kansas Pacific Railroad to lay a siding at the hamlet of Abilene, Kansas, on the edge of the quarantine area. Then he persuaded Kansas not to enforce the quarantine in this area of the state. Kansas offered access to the railroad and a grassy passage to central Texas. McCoy began building pens and loading facilities and sent word to Texas cowmen that a cattle market was available. That year he shipped thirty-five thousand head; the number doubled each year until 1871, when six hundred thousand head glutted the market.[7]

The first herd to follow the future Chisholm Trail to Abilene belonged to O. W. Wheeler. In 1867, Wheeler had purchased twenty-four hundred steers in San Antonio. He planned to winter the cattle on the plains and then move them by trail to California, where a $14 steer could garner more than $100 in the gold fields. They followed wagon tracks and wound up on a trail made by Scot-Cherokee Jesse Chisholm, who in 1864 began hauling trade goods to Indian camps from his camp in Wichita, Kansas. Normally, Texas herds followed the old Shawnee Trail from San Antonio, Austin, and Waco to the north. The Chisholm Trail continued to Fort Worth, then passed east to cross the Red River to Abilene.

Thanks to the rich prairie grasses, when conditions were favorable, cattle actually gained weight on the trail. A trail boss, ten cowboys, a cook, and a horse wrangler could move twenty-five hundred cattle for three months at a cost of sixty to seventy-five cents a head. This was the era of the cowboy—although not the romanticized experiences seen in film and fiction. Most often, the drudgery of moving cattle was punctuated with only rare violent weather or stampedes. Cattle drovers protected herds from rustling, which was carried out by Indian and Mexican groups, but most often by disgruntled trail hands or cowboys. To make rustling more difficult, many ranchers began employing brands (which had first been used by Spanish mission ranches). Unique to each ranch, brands were burned into the flesh of each steer and could not be removed. Branding cattle brought some order to cattle on the open range and made the animal's meat a commodity in the larger economic system. At its most basic level, though, the branded steer represented a tool for making it profitable to own some of the vast, open space of the American West.

Annihilating Space in the United States: The Grain Elevator

In the 1850s, many Midwestern cities began utilizing a device found on many farms that grew grain: the silo. The urban model, however, was much larger and became known as an elevator. Intimately linked into the system of growing grain and transporting it by railroad, the elevator became a revolutionary device when used to store grain for market-driven reasons. Historian William Cronon writes: "Chicagoans began to discover that a grain elevator had much in common with a bank—albeit a bank that paid no interest to its depositors."[8] Primarily, the elevator stored the grain and helped growers maintain a supply throughout the year. It functioned as kind of a holding tank.

In regions such as the American West, grain elevators helped to create a systematic infrastructure for grain storage all along the railroad. Such a system made far-flung agriculture profitable and reliable. LIBRARY OF CONGRESS PRINTS AND PHOTOGRAPHS DIVISION

As if carrying a precious mineral, farmers brought their wheat or corn to the elevator operator. The operator gave the grower a receipt that could be redeemed for grain when the original grower wished to make a withdrawal. Early on, grain was measured, traded, and sold in sacks. In order to simplify its trading, grain operators adopted a "liquid" form of measuring it that liberated the grain from the sack. This was the change that allowed the elevator to take over the landscape of each Midwestern city.

In cities such as Chicago, the supply of grain seemed endless. Supplying much of the nation's needs, the Chicago elevator operators were grain brokers. Much as a bank, the brokers bought and sold grain. The elevator, though, allowed this financial market to become a "futures" market. Cronon writes: "Grain elevators and grading systems had helped transmute wheat and corn into monetary abstractions, but the futures contract extended the abstraction by liberating the grain trade itself from the very process which once defined it: the exchange of physical grain."[9] Although making possible the massive growth of an industry, the elevators assured, again, that the open spaces of the West could provide reliable profit for farmers. Powered by coal, railroads made the entire system function.

RADICAL ADAPTATION AT THE COLUMBIAN EXPOSITION

At the basic level of the formation of this epoch in human living is an acceptance of ideas or systems that organized advancements in technology in ways that ultimately changed foundational aspects of everyday existence. Most relied on an unflinching supply of power. Such radical technological changes were remarkable; even more transformative, though, was societies' capability to deliver those new technologies through innovations such as the railroad directly into consumers' lives and homes, thereby wildly multiplying their impact. From the Crystal Palace in 1851, to the Centennial International Exhibition in 1876 (Philadelphia), and on to the Columbian Exposition in 1893 (Chicago), new ideas were celebrated and technological innovations presented as progressive on a transborder scale that was previously impossible. More than in any previous era, a collective vision of human progress was attainable if certain systems and technologies could be acquired.

In 1893, the United States rallied as never before to direct its combined resources of finance and technology toward a single end that showcased the system that railroads had enabled. The outcome was not a new method of producing goods, a remedy for a common illness, or a weapon to defeat a desperate enemy. And, yet, the outcomes eventually included these and many other great accomplishments of the twentieth century. Similar to much of twentieth-century

American life, these outcomes derive from the modern idea of using technology to solve problems of everyday, domestic life. This was the year that Americans concentrated their abilities and their aspirations to create a global spectacle in the World's Columbian Exposition in Chicago. Visitors, it was hoped, would tour the exposition and come away with an understanding of where the nation intended to go in the future. The designers hoped that by creating the site they would in some small way help to create that future. To develop this site, designers manipulated nature to make a park and fairground that could thrill and impress any visitor. In doing so, the creators of the Columbian Exposition produced a symbol for an age when technology and the solutions of engineering were no longer relegated to factories and workplaces. In an apt representation of the emerging modern society, behind the scenes fifteen steam engines drove the exposition. These prime movers could create 13,000 horsepower to drive sixteen generators to produce 8.955 kilowatts of power, which was sufficient enough to light 172,000 incandescent bulbs, including the largest searchlight in the world, which weighed six thousand pounds, and an eight-foot-tall light bulb.[10] Historian David F. Burg writes that nearly everything "that glowed, that sounded, that moved at the fair was powered by electricity."[11]

Historian Thomas Hughes calls technology the "effort to organize the world for problem solving so that goods and services can be invented, developed, produced, and used."[12] The surge of American society toward technology grew out of European models of economic development. Hughes continues:

> The Europeans held up a mirror in which the Americans could see themselves as the raw materials of modernity which the Europeans wanted to fashion into modern culture. European engineers, industrialists, artists, and architects came to America to admire its "plumbing and its bridges" and made . . . the second discovery of America—the great systems of production. . . . In so doing they were inventing the forms and symbols for a modern technological culture.[13]

This rethinking of the human condition, including its relationship to the natural world, helped to make the twentieth century one of dramatic change. Historian Stephen Kern writes: "These old scaffoldings had supported the way of life and culture of the Western world for so long that no one could recall exactly how they all started or why they were still there, and it took a generation of restless scientists, artists, and philosophers to dismantle them."[14] In Kern's argument, the modern sensibility redefined essential human concepts, including those of time, space, and self. In each of these relationships, technology now brokered an important role and when wound with ideas of energy management, planning, and regulation, it reconfigured entire societies.

CONDUIT: The Standard Oil Kerosene Can

The building material stood out distinctly against the huts of natural materials that predominated the Far East: blue tin stacked, propped, or stuck to other organic matter. Throughout portions of China, India, and areas such as Burma and Sumatra, humans lived very simply into the late 1800s. Obviously, this new building material was an interloper, imported from a far-off land across space and, truly, time. Used only once and then discarded, the cans came from a distant country with manufacturing prowess that took valuable raw materials such as metal and fabricated from them disposable items. The cans had found their way into locales in which most cooking occurred over a fire and education was limited. The presence of so few manufactured goods in such places meant that any waste materials might find new uses, such as serving as building supplies. If one looked closely at the torn tin one might also make out the scripted words: "Standard Oil of California."

The use of personal combustion devices such as the Lucifer match joined with basic lamp technologies by the mid-1800s to make lighting a consumer product. Typical histories might just gloss over such a development; telling the story of energy, though, requires that *THHN* notes the moment that brought ubiquitous—even thoughtless—illumination to much of the world by the 1900s. The flexible lamps of the nineteenth century allowed consumers to utilize a variety of oils and, therefore, the marketplace that began with whale and sperm oil became the competitive marketplace defined by mineral-based oils that could conceivably reach consumers all over the world in a tanker from which the liquid energy could then be put in a portable can. Most often, the kerosene that such cans contained provided fuel for heating, cooking, or lighting. Such uses, though, of course, still required the use of a match to combust the liquid.

The fuel in the can was an American original. Although it was known in other parts of the world for various purposes, petroleum's synchronic relationship with the American spirit of expansion began in 1859 and has never entirely gone away. Patterns of boom and bust entirely defined the early industry, however, and as the development of crude moved throughout the globe, corporate and national leaders realized the great benefit of organizing capital around centralized planning and systemization. It is ironic that the primary example for this international model of development sprang from the American experience with crude represented by the tin can stuffed into the wall of a hut far from the well from which the oil had sprung.

By reacting against the unpredictable cycle of boom and bust, Rockefeller's Standard Oil Company was a model for other nations (and was also their competitor) as they devised ways to prosper from their own supplies of crude. In other regions, nations developed the resource themselves or joined with existing petroleum companies (including Standard). Typically, enterprises following the international model relied much less on independent speculators and wildcatters. Often, the large systems around which these international enterprises were organized were constructed from necessity: lacking their own supply of petroleum, a few powerful nations sought the technical and political mechanisms to acquire petroleum wherever it occurred. In doing so, the British, Dutch, and French foreshadowed an era of petroleum scarcity that would befall all nations in the twenty-first century. By the late 1800s, petroleum was simply another resource organized into the mercantilist tradition that had become known as colonialism long before the early twentieth century. The international model, therefore, most often was carried out by European oil firms in cooperation with a crown or government that exerted colonial authority in the oil-producing region.

Although little changed in terms of petroleum's usefulness by the late 1800s, new technologies helped to tie together a highly organized, international model of petroleum development by prioritizing methods for dispersing kerosene all over the world—no matter where the petroleum from which it derived had been harvested. Although each aspect of harvesting and dispersing crude required significant economic capital, techniques such as pipelines brought crude to centralized processing and refining locations. And tanker railroad cars allowed refined oil to be moved more easily to points of use or shipment. In these global examples, once supply was reliably established, this expanding infrastructure of dispersal became the next hurdle and the next frontier of great opportunity and profit for creating the high-energy existence. While its dispersal relied on an elaborate, complex corporate organization, the humble can served as a conduit for the commodity while also bridging lifestyles and civilizations.[15] It was a very early example of a truly portable, flexible prime mover that could be delivered directly to consumers all over the world. But it did require a Lucifer match!

OCEAN LINERS ALTER TIME, SPACE, AND TRADE ON A GLOBAL SCALE

New transportation technologies also created systems for economic development that extended those initiated during the Age of Sail. For instance, the Standard

Oil can mentioned above was made possible after the first shipment of kerosene left Philadelphia for London in 1861.[16] Although there was great fear on board about the cargo's safety, once the trip proved uneventful and the kerosene remarkably useful, Europeans clamored for more and word spread to Russia and elsewhere of the "new light." Through the 1860s and 1870s, most would come through the hands—and tin cans—of Standard Oil and originate from the fields of Pennsylvania. By the 1890s, international competition to develop and ship petroleum all over the globe led Marcus Samuel, a London merchant with shipping connections throughout the Far East who worked for Standard's competitors, to send out a fleet of oil tankers to pass through the Suez Canal in 1891. He assumed that global consumers would have empty cans that could be filled. Samuel did not realize that Standard's tin cans had in their own right become prizes of people throughout the Far East.[17] Adjusting quickly to maintain the competitive advantage that he had earned for the Rothschilds, Samuel immediately sent ships laden with raw tin to be fabricated into vessels. His red cans were also soon additionally being used in home construction and other activities completely unrelated to the oil that they had once carried. Clearly, it was a moment of worlds meeting across time and space.

On the global scale, coal and later petroleum combined with new marine technologies to lead humans from the Age of Sail and into a more global society in which ocean liners redefined ideas of time and space and helped to forge new connections throughout the world. Steamships had become more and more prominent for intercoastal trade after 1860, but the shift among oceangoing vessels was somewhat slower. Petroleum allowed oceangoing vessels to shift to steam more rapidly by the early twentieth century.[18] The great trans-Atlantic liners were primarily made of iron and steel by British ports. These liners became crucial devices for remaking the nature of human ideas of time, space, and connectivity by standardizing travel throughout the world.

Most important for patterns of human movement, steerage tickets made it possible for laboring-class immigrants to move more flexibly throughout the world. This was particularly evident as the "golden door" of migration fueled American industrial growth, making it one of the most attractive destinations for those in search of work. The early 1900s, therefore, became the period of greatest human migration in history with massive numbers of Europeans arriving via New York City and other major ports. The sheer mass of immigrant arrivals led to the creation of American symbols such as the Statue of Liberty, designed to welcome the poor and hungry of other nations, and, of course, Ellis Island to organize and classify each new immigrant. These icons were an important part of systematizing the global movement of humans after 1900.

ELECTRICITY AND THE EVOLUTION
OF THE ENERGY INDUSTRY

Humans accepted their ingenuity and its ability to remake their place in the world. Through engineering, Americans particularly defied the limitations of previous eras. Both fueling and helping to cause this transition were new ways of living with energy. For instance, electricity is not an energy source; it must be created by extracting power from another prime mover and converting and transferring it for use elsewhere (or storing it in some fashion to save it for future use). This flexible transfer of power made electricity the single most important technology of the new era and also brought entirely new avenues through which energy could be used in human life. Indeed, in its most revolutionary form, electricity took the power of coal that had remade human work and industry and domesticated it.

In the nineteenth century, a growing abundance of energy remade industry and work; however, technologies for basic domestic applications such as illumination remained similar to the past. Innovators, including American luminaries such as Benjamin Franklin, experimented with aspects of capturing natural forms of electricity. Beginning in 1743, Franklin worked with a London friend, Peter Collinson, to use a glass tube for generating static electricity, which was the only type known at that time. Franklin wrote of hosting an "electric picnic" along the Schuylkill River in 1749 where his group ignited flammable spirits on one bank of the river and then sent the spark from side to side through the river, using the water as a conductor.[19] As part of the outing, Franklin's picnic would use a series of Leyden jars, one of the earliest batteries, to store the charge briefly (and then use the power to cook a turkey!). While chemical batteries had not been devised, Franklin used the Leyden jars during his famous 1752 experiment with lightning. For these experiments, the Royal Society of London awarded him its Copley Medal in 1753.

The effort to better store an electric charge (and thereby harness it for use in other applications) was carried out in 1800 by Luigi Galvani, an Italian surgeon and physiologist, and an Italian inventor, Alessandro Volta. Through a long series of experiments, they transferred electric charges to contract the muscles of frog legs.[20] Volta stacked disks of metals, including copper or silver, tin or zinc and separated them with saltwater-saturated cardboard in an instrument that he called a *pila* (pile). Many other innovators, then, expanded these essential ideas into a usable battery that might be used to store an electric charge generated from fossil fuel (coal, oil, natural gas), hydroelectric (water power), or, eventually, nuclear power. As the scale of the undertaking grew, the electric utilities industry came to include a large and complex distribution system and, as such, is divided into transmission and distribution.

Following these experiments in Europe, the next challenge was electrical transmission, and this fell, in particular, to the mind of Thomas Edison, one of the America's greatest inventors. In 1878 Joseph Swan, a British scientist, invented the incandescent filament lamp and within twelve months Edison made a similar discovery in America. Edison used his direct current (DC) generator to provide electricity to light his laboratory and later to illuminate the first New York street to be lit by electric lamps, in September 1882. From this point, George Westinghouse patented a motor for generating alternating current. Society became convinced that its future lay with alternating current (AC) generation. This, of course, required a level of infrastructural development that would enable the utility industry to have a dominant role over American life.

Once again, this need for infrastructural development also created a great business opportunity. Samuel Insull went straight to the source of electric technology and ascertained the business connections that would be necessary for its development. In 1870, Insull became a secretary for George A. Gouraud, one of Thomas Edison's agents in England. Then, he came to the United States in 1881 at age twenty-two to be Edison's personal secretary.[21] By 1889 Insull became vice president of Edison General Electric Company in Schenectady, New York. When financier J. P. Morgan took over Edison's power companies in 1892, Insull was sent west to Chicago to become president of the struggling Chicago Edison Company. Under Insull's direction Chicago Edison bought out all its competitors for a modest amount after the Panic of 1893. He then constructed a large central power plant along the Chicago River at Harrison Street. The modest steam-powered, electricity-generating operation would serve as Insull's springboard to a vast industrial power base.

By 1908, Insull's Commonwealth Edison Company made and distributed all of Chicago's power. Insull connected electricity with the concept of energy and also diversified into supplying gas. Then he pioneered the construction of systems of dispersing these energy sources into the countryside. The energy grid was born. It would prove to be the infrastructure behind each human life living in the developed world in the twentieth century. Through the application of this new technology, humans now could defy the limits of the Sun and season; but even more important, industrial productivity of all sorts could expand exponentially.[22] On the whole, new energy made from fossil fuels altered almost every American's life by 1900. In 1860 there were fewer than a million and a half factory workers in the country; by 1920 there were eight and a half million. In 1860 there were about 31,000 miles of railroad in the United States; by 1915, there were nearly 250,000 miles. In addition, the energy moving through such infrastructure would not remain limited to the workplace.

CONDUIT: Edison's Light Bulb

For decades, inventors and businessmen had realized that the great leap for democratizing the use of electricity would be to make it a source of illumination—essentially, to make the match unneeded to illuminate the lamp. Unlike any of the oils or gases being used, it would involve no open flame, no need for lighting, and no need to refill. In the late 1800s, electric lighting was a dream whose convenience made it worth pursuing. However, the technical challenges were quite overwhelming. While some innovators emphasized the system that would be necessary to transmit electric current, other inventors focused on making a bulb that would receive the current: primarily, they experimented with positioning a filament in a vacuum tube. The electric current, then, was sent through in hopes of making the filament glow. The filaments consistently failed, though, disintegrating as soon as the current reached them.[23]

In 1878, Edison decided to concentrate his inventive resources on perfecting the light bulb. Instead of making his filament from carbon, Edison switched to platinum, which was a more resilient material. In 1879 he obtained an improved vacuum pump called the Sprengel vacuum, and it proved to be the catalyst for a breakthrough. Using the new pump, Edison switched back to the less expensive carbon filaments. Using a carbonized piece of sewing thread as a filament in late October, Edison's lamp lit and continued to burn for thirteen and a half hours. Edison later changed to a horseshoe-shaped filament, which burned for over a hundred hours. Edison had invented a practical light bulb; more important, he cleared the path for the establishment of the electrical power system that would revolutionize human existence.

It was this power system that became Edison's real achievement and created the market that would beget a huge new industry destined to alter the lives of every human. The nature of everyday life became defined by activities made possible by electric lighting as well as the nearly endless amount of other electrically powered items. The light bulb was a critical innovation in the electrification of America; however, it also helped to create the market that stimulated efforts to perfect the industry of power generation.[24] At the root of power generation was the dynamo. The dynamo was the device that turned mechanical energy of any type into electrical power.[25] When Edison started working on the light bulb the most effective dynamo produced electricity at approximately 40 percent of the possible efficiency. He developed a dynamo that raised this efficiency to 82 percent.

Together, these technological developments made it possible for Edison to start providing electricity commercially to New York City. By September

1882 he had opened a central station on Pearl Street in Manhattan and was eventually supplying electricity to a one-mile-square section of New York. These areas—much like the transcontinental railroad and the Brooklyn Bridge—became futuristic symbols for the growing nation and set aspirations for humans everywhere.

HYDROELECTRIC DAMS AND REGIONAL POWER DEVELOPMENT

By the early 1900s, the desire for electricity joined with engineers' ability to modernize one of the first prime movers in humans' past: water power directed through a dam. While the technology of river management through dams remained a global ingenuity for managing flood control and irrigating dry regions, American experiments with a flexible dynamo allowed dams to also serve as generators of electricity. Freestanding hydroelectric dams began appearing throughout the United States by the late 1800s. Very quickly, the system-builder mentality had grasped these dams' capability to transform entire regions or nations, particularly in areas without easy connection to an existing electrical grid. Hydroelectric dams became the key design elements to some of the largest scale planning in human history—with each project organized by the desire to create a self-contained supply of the electricity that was rapidly proving capable of transforming societies.

In Egypt, dreams of electrification grew out of Britain's success at using the Aswan Dam construction to manage the Nile River's flow for agriculture. In such an example, the technology to build dams and electrical systems became an important export during the modern era of colonialism. Dams might offer less-developed nations a pathway to almost immediate modernization with electricity at its center. Britain occupied Egypt beginning in 1882 and sought to use its advanced technology to assist Egypt's water supply. A low Aswan Dam was erected in 1902 and then heightened in 1912 and 1934. While the interest in adding power-generating capabilities began in the 1800s, the Aswan High Dam was not accomplished until the 1970s as a national modernization project (it will be discussed in a later chapter).

Southern Europe, however, spurred significant growth through hydroelectricity in the Po Valley after 1890. Known as "white coal," hydroelectric power development in Italy brought significant industrial development to Milan and Turin. With scarce supplies of actual coal, the Italians sought electrification from the Alpine torrents that fed the Po through a series of dam projects. Completing

Hydroelectric dams such as this one on the Elwha River in
Washington turned the flow of the river into electricity by
running it through a dynamo. LIBRARY OF CONGRESS PRINTS
AND PHOTOGRAPHS DIVISION

its first big dam in 1898, Italy led all of Europe in hydro use by 1905 and by
1937 it furnished almost all of Italy with electricity. Indeed, with this power
source in place, Milan was the second city in the world to electrify its street lighting. It is not overstatement that Italy's emergence as a European and imperial
power after 1890 grew out of this electrification based in the Po Valley.[26]

Although the United States encompassed the breadth of the North American
continent, regional developments throughout also grew out of large-scale hydroelectric planning. Along the Colorado River, settlement of the American
West was tied to a system of large dams, including Boulder (Hoover) Dam, that
allowed cities such as Las Vegas, Nevada, Los Angeles, California, and Phoenix,

Arizona, to grow into urban centers with more than fifteen million residents depending on its electricity. The New Deal of the 1930s spurred regionally specific hydroelectric projects in the Columbia and Tennessee River valleys. Particularly in the case of the Tennessee, this rural electrification brought modernization to one of the nation's least progressive regions. The power played a critical role in weapons production during World War II in both regions.

MECHANIZATION EXPANDS AGRICULTURE

A far cry from methods of the biological old regime, agriculture also mechanized in a number of important ways that catapulted production globally—despite the fact that by 1900 most arable land was already in production. During the first era of industrialization, agriculture in Europe and elsewhere was greatly altered by transportation innovations. In Western Europe and the United States, railroads helped to create an infrastructure for food creation and dispersal. Similarly, steam shipping allowed massive amounts of grain to begin moving by sea throughout the world. And this early phase was just the start of such changes.

Founded on the bountiful petroleum supplies, agricultural production increases grew from advancements in chemical fertilizers and pesticides, irrigation, agricultural machinery, and plant breeding. While farming in 1900 was carried out quite similarly to how it had been done for thousands of years, new tools used fossil fuels to increase capabilities, particularly in the United States, Soviet Union, Canada, Argentina, and Australia. In the period when agriculture in developed countries derived its energy form fossil fuels "in the form of diesel oil and petrol, rather than fresh hydrocarbons such as hay and oats," writes economic historian Christian Smedshaug, "large areas of land were released."[27] Mechanization, such as tractors and other equipment, allowed larger areas to be farmed but they also enabled the transfer of land reserved for feeding work animals. The year 1915 marked the high point for the use of draught animals.[28] Gasoline-powered tractors allowed fewer farmers to tend to more and more land. Conversion from animal power began in the United States after World War I and the USSR in the 1930s. Harvesting combines encouraged farmers to veer toward monoculture (growing one crop) but vastly expanded the land that could be tended. And such devices also meant that feed crops for work animals were no longer needed. Across the board, modern farming was defined by increased systemization and reliability.

Joining with other forces of modern agriculture, this expansion had allowed industrialized nations to produce significant food surpluses by the late nineteenth century that were further compounded by other applications of petroleum after 1920.[29] Crop reliability was also enhanced by new understandings and capabilities of fertilizers. By 1900 natural fertilizers such as guano had been used up and essential manufacturing materials such as saltpeter were reserved for the preparation of munitions. The synthesis of ammonia and the consequent industrial production of nitrogenous fertilizer freed growth of the human population from the natural limits of the biological old regime.

Often using petroleum hydrocarbons as a basis or active ingredient, chemicals became available to control some of the problems associated with agriculture, including insects and weeds. A variety of chemicals, each derived from or using the energy of petroleum, has been used by farmers to either stimulate growth or control pests, particularly herbicides and pesticides. The petrochemical industry emerged to manufacture these products during the mid-1900s in massive factories that resembled gigantic laboratories. Although any refinery is needed to make oil into each fuel product, these manufacturers took the chemistry further and used the oil as the basis for chemicals that became essential to agriculture in developed nations. In most cases, the petrochemical industry also became one of the world's greatest polluters, leaving toxic residue behind wherever such chemicals were manufactured. Most important to twentieth-century agriculture, though, chemists also discovered methods for creating synthetic nitrogen to enhance growing potential—a process that also required petroleum.

Although Earth's supply of nitrogen is limited, all life depends on it to serve as the building block from which amino acids, proteins, and nucleic acids are assembled. Plants do not grow without nitrogen, and most of Earth's useable supply could be found stuck to the roots of leguminous plants. Synthetic nitrogen derives from chemists' effort to "fix" it, which is a laboratory process of applying heat that splits nitrogen atoms and then joins them to hydrogen in order to create synthetic replicas. In 1909, Fritz Haber and Carl Bosch created a method for nitrogen fixing that was largely responsible for the manufacture of synthetic fertilizer during the twentieth century.

The Haber–Bosch process combines nitrogen and hydrogen gases under intense heat and pressure, which is supplied by electricity, and the hydrogen is supplied by fossil fuels, normally oil, coal, or natural gas. Journalist Michael Pollan writes: "When humankind acquired the power to fix nitrogen, the basis of soil fertility shifted from a total reliance on the energy of the sun to a new reliance on fossil fuels."[30] The supply of nitrogen fertilizer was quite suddenly nearly boundless, thus liberating humans from the constraints of natural limits and allowing farms (particularly those in the United States) to be managed under industrial principles. Pollan adds: "Fixing nitrogen allowed the food chain to turn from the logic of biology and embrace the logic of Industry. Instead of eating exclusively from the sun, humanity now began to sip petroleum."[31] What became known as the Haber–Bosch process shaped subsequent world history by lifting some of the natural limitations on plant growth and thereby vastly lifting the human potential to feed itself. As with other examples of new technologies, though, increased food production most benefited developed nations, thereby adding considerably to the emerging gap.[32] With this artificial boost by applying energy, agriculture after 1950 could expand wildly beyond the natural constraints of the land.

CURRENT: Modernism and European Innovations Create Global Progress

Whether in agriculture or transportation, at the base of the new way of living in consumer nations were applications of new, ubiquitous forms of energy. New machines and know-how seemed to put each human firmly in control of creating their own world, which might work much better than the one that naturally occurred around them. An extreme version of this view took root in Europe and became known as modernism. Particularly in terms of the materials used for building, modern structures prioritized elements created by new processes. However, in both their form and function, modern structures integrated the capabilities of new technologies that used electricity to allow homes and buildings to function more effectively than those of previous generations.

Sometimes referred to as white architecture, this style is recognizable for its flat painted surfaces, box-like dimensions, and metal-framed windows that sit flush against the facade. International-style buildings are made with modern materials and contain no ornamentation or decorative flourishes. Architects often repeated shapes on the surface of a building (to affect the look of a paper-thin screen) to diminish the mass of the structure. The idea was to create the impression of space closed in by thin walls. These artificial environments often used a great deal of glass to allow for natural lighting and the impression of a wall-less space. The style, which was officially named "The International Style" after a 1932 exhibition at the Museum of Modern Art in New York City, was powered by Walter Gropius, the German architect and founder of the Bauhaus School, and, eventually Ludwig Mies van der Rohe and Walter Gropius, the Dutch architect J. J .P. Oud, and the French pair of Le Corbusier and Pierre Jeanneret.

Referring back to essential elements used in the Crystal Palace discussed above, the works of these modern designers informed a new generation of buildings that changed the human living environment during the twentieth century. At the core of these structures, building technologies became simpler and more flexible. Of course, this additionally meant that building could take place at an increased rate and scale.

STEEL SKELETONS AND THE SKYSCRAPER

By integrating new technologies, designers entertained the idea of concentrating human activities by making cities grow upward as well as outward. Whereas churches with bell towers or steeples were often the tallest structures in urban landscapes during the 1800s, office buildings grew taller by using iron and steel to create an internal skeleton. This change was particularly evident in American cities, primarily Chicago and New York. For instance, in 1889, the tallest building in the United States was New York's Trinity Church, near Wall Street. The next year, it was overtaken by the twenty-six-story New York World Building. Fueled by an intense demand for office space in rapidly expanding downtown areas, the skyscraper became a symbol of the modern era of industrial development.[33]

Products of massive energy supplies, technological innovation allowed architects to overcome the limits of gravity. Brick could not bear the weight of buildings higher than five or six stories; however, beginning in Chicago in 1884, steel frame construction allowed architects to design buildings of unprecedented height. The other essential technologies included elevators and a way of moving heat around the tall, narrow space. Typically, these early skyscrapers rose approximately twice as high as neighboring structures.

William Le Baron Jenney, a Chicago architect, is credited with designing the first skyscraper in 1884. Nine stories high, the Home Life Insurance Building was the first structure whose entire weight, including the exterior walls, was supported on an iron frame. But it would not be for another fourteen years, when the Equitable Life Assurance Building was constructed in Manhattan, that a skyscraper contained all the characteristics of a modern skyscraper, including central heating, elevators, and pressurized plumbing.[34] Although it was possible to construct buildings more than sixteen stories high using masonry walls to support the weight, the buildings had to have such thick walls and small windows that they were unappealing to landlords. The falling price of steel during the 1880s meant that tall buildings with steel frames became cheaper to build. The metal skeleton supported not only the roof and floors but also the external walls.

To transport people within the building, skyscrapers needed elevators. During the 1870s, some five- and six-story buildings had steam-powered elevators, which had cables wound around a huge rotating drum. Taller buildings demanded a different technology, because the drum would have to be impractically large. When it was built in Paris in 1889, the Eiffel Tower used hydraulic-powered elevators, which required a huge power source. During the 1890s, the electric elevator offered a more practical solution.

Tall buildings also needed ventilation systems to heat them in the winter and cool them in the summer. The early ventilation systems, introduced in the 1860s, used steam-powered fans to move air through ducts. After 1890, fans

Coney Island with its electric-powered attractions helped to entice humans to the high-energy existence of the twentieth century. LIBRARY OF CONGRESS PRINTS AND PHOTOGRAPHS DIVISION

were driven by electricity. Steam heating using radiators was widely used by 1885. Plumbing to circulate water through the building relied on pressure using electric pumps.

As these early examples succeeded and the technology refined by 1930, skyscrapers grew—literally—and also spread to other nations. The greatest symbol of this era, of course, was New York City's Empire State Building, which formally opened on May 1, 1931. President Herbert Hoover and New York governor Franklin D. Roosevelt attended the dedication of the 102-story, 1,250-foot high building. Erected in just thirteen months, the building grew at a rate of more than one story per day. Construction workers balanced on girders a fifth of a mile above the ground. The completed building seemed to defy the nature of the human need to remain on Earth.

With the use of these new technologies, skyscrapers became a symbol for all modern technology and its ability to alter human life. By integrating energy into methods of construction as well as into the structure itself, these structures relieved urban congestion and allowed population and workplace concentration that had never been possible before. Particularly in the United States, technological innovation left very little unchanged in human life by the 1910s and the skyscraper rose as one of the great symbols of the modern machine age, made possible with ubiquitous and inexpensive power.[35]

REWORKING HUMAN CULTURE AT CONEY

It was only a matter of time until the increasing mechanization of industry began to creep into other parts of everyday life to also change what humans found
interesting, thrilling, or amusing. The modern era found its version of the bazaar
or marketplace in a place such as Coney Island, New York. Through immigra-
tion, factory growth, and skyscrapers, New York City took a defining global role
in establishing life in the modern era by the early 1900s. While technology and
power had increased production and consumption in many parts of the world, in
the United States it also defined this new era by linking humans to the way they
lived outside of the workplace.[36] At Coney Island, ubiquitous electricity created
a new world of entertainment and escape.

Coney Island proved that this new era of technology could link together diverse
humans. New, modern ways of doing things helped to create portions of culture
that were designed to appeal to masses, which often helped people to overcome
differences of race, gender, ethnicity, and economic class. It was at places such as
Coney Island that the nature of human contact with itself and the world around
it changed dramatically. The new era of electricity helped to create this cultural
commons, powering the attractions and bringing many visitors from throughout
New York City when the subway system arrived in the early 1900s. The resort
attracted a hundred thousand visitors on summer Sundays in 1900 and five hundred thousand daily during the 1910s and a million per day by the 1920s. The
ocean remained a primary attraction to visitors; however, the landscape of leisure
that took shape included many additional elements as well, including Luna Park,
Steeplechase Park, and a host of other attractions and food concessions.[37]

Through modern technology, every day was a festival day—such as Oktoberfest
in Germany. New technologies (each drawing on the power of electric currents)
applied modern sensibilities to find common ways to thrill any human. The
new machines of the early twentieth century altered the most basic aspects of
the human's involvement in nature. Revolutions rippled through human limits
including speed, gravity, and even ideas of fun. The basic definition of the thrill
ride was that it must strain the laws of human existence. The Ferris wheel, of
course, was about rising into the sky. More often, though, the thrills came from
new speeds with which the human body could be flung through the air.

Historian Judith Adams writes that early Coney developers created rides and
thrills that altered visitors. "Their muscles were loosened and their inhibitions
shattered mostly by simple mechanical contrivances designed to strip visitors
of all means of control."[38] In the nineteenth century, opportunities for escape
relied on individual imagination, such as in literature, art, and story-telling. In
the twentieth century, technology became a crucial new element to creating momentary escape and diversion. Technological diversions such as those at Coney

helped to create mass culture. Drawing energy from electricity, later innovations continued this development with film, television, and other media as well.

PETROLEUM REVOLUTIONIZES HUMAN MOVEMENT

The intricacy of petroleum to human life today would have shocked nineteenth-century users of "Pennsylvania rock oil." Most farmers who knew about the oil in the early 1800s considered it a nuisance to agriculture and water supplies. These observers were not the first people to consider the usefulness of petroleum, which had been a part of human society for thousands of years. Its value grew only when European Americans offered the resource their commodity-making skills.

Throughout the biological old regime, crude oil was found and used in some fashion in various locales throughout the world. However, the area that is credited with first noticing petroleum is a mountainous area in western Pennsylvania, nearly one hundred miles above Pittsburgh. The oil occurring along Oil Creek was named initially for the Seneca people, who were the native inhabitants of this portion of North America at the time of European settlement. Paleo-Indians of the Woodland period, before 1400, also came to the area to harvest the oil used in their religious rituals.[39]

Europeans began bottling the loose crude in the 1840s and selling it as a mysterious cure-all that had nothing to do with its capabilities as a source of energy. Developers of oil for illumination began experiments with petroleum in the 1850s. In 1857, the Pennsylvania Rock Oil Company of Connecticut sent Edwin Drake to Pennsylvania to attempt to drill the first well intended for oil. The novelty of the project soon had worn off for Drake and his assistant Billy Smith. The townspeople irreverently heckled the endeavor of a "lunatic." During the late summer of 1859, Drake ran out of funds and wired to New Haven, Connecticut, for more money. He was told that he would be given money only for a trip home—that the Seneca Oil Company, as the group was now called, was done supporting him in this folly. Drake took out a personal line of credit to continue, and a few days later, on August 29, 1859, Drake and his assistant discovered oozing oil.[40]

After the American Civil War, the industry consistently moved toward the streamlined state that would allow it to grow into the world's major source of energy and lubrication during the twentieth century. Oil was a commodity with so much potential that it attracted the eye and interest of one of the most effective businessmen in history, John D. Rockefeller Sr. Working within the South Improvement Company for much of the late 1860s, Rockefeller laid the groundwork for his effort to control the entire industry at each step in its process. Rockefeller formed the Standard Oil Company of Ohio in 1870. Oil exploration grew from the Oil Creek area of Pennsylvania in the early 1870s and would expand from Pennsylvania to other states and nations during the next decade. By 1879 Standard controlled 90 percent of the U.S. refining capacity, most of the rail lines between urban centers in the northeast, and many of the leasing companies at the various sites of oil speculation. Through Rockefeller's efforts and that of the organization he made possible, petroleum became the primary energy source for illumination for the nation and the world, ultimately filling the cans such as that discussed above.

POWERING THE FIRST AUTOMOBILES

In one of the most consequential shifts to our species, massive supplies of energy also dramatically altered patterns of human movement at the start of the twentieth century. The automobile is part of the individualization of transport that can be traced as far back as the fifteenth century when the Renaissance genius

Hitting the road became a passion for humans even before towns and cities were well connected to one another. LIBRARY OF CONGRESS PRINTS AND PHOTOGRAPHS DIVISION

Leonardo da Vinci considered the concept of a self-propelled vehicle and Robert Valturio planned for a cart powered by windmills geared to its wheels. As early as the sixteenth century, while animal- or human-powered transportation was the norm, steam propulsion was proposed as an alternative prime mover by inventors all over the world. In 1678, a Belgian missionary to China, Ferdinand Verbiest, made a model steam carriage based on a principle that suggests the modern turbine. In the seventeenth century the great Dutch physicist Christiaan Huygens built an engine that worked by air pressure developed by exploding a powder charge. A carriage propelled by a large clockwork engine was demonstrated around 1750 by the French inventor Jacques de Vaucanson.[41]

The first invention to clearly foreshadow the automobile of the early twentieth century was the three-wheeled steam-powered carriage. Steam carriages were produced in England during the late eighteenth and early nineteenth centuries through the work of William Murdock, Richard Trevithick, Goldsworthy Gurney, and Walter Hancock. Hancock's "steam bus," built in 1832, was in regular service between London and Paddington. In the United States, Oliver Evans in 1805 built the first steam-powered motor vehicle that operated on land and water. Richard Dudgeon's road engine of 1867, which resembled a farm tractor, could carry ten passengers. Steam-driven automobiles were turned out by some one hundred manufacturers during the late 1890s and early 1900s. The most famous of these steam-car makers were Francis E. and Freelan O. Stanley of the United States—twin brothers who developed an automobile called the "Stanley Steamer" in 1897. Steam cars burned kerosene to heat water in a tank that was part of the car. The pressure of escaping steam activated the car's driving mechanism. The popularity of the steam car declined at about the time of World War I and production came to an end in 1929. The desire for a way of modernizing human movement swept the developed world; however, a lack of reliability kept inventors searching.

Electricity was a clear leader to possibly power transportation devices, and during the 1880s a number of electrically powered automobiles were built in Europe. One of the first "electrics" in the United States was produced by William Morrison in 1891 and the market grew quickly. Between 1896 and 1915, their high point in popularity until recently, fifty-four U.S. manufacturers turned out almost thirty-five thousand electric cars. The Columbia, the Baker, and the Riker were among the more famous makes. The electric car ran smoothly and was simple to operate; however, it did not run efficiently at speeds of more than twenty miles per hour and could not travel more than fifty miles without having its batteries recharged. Thus, it was limited to city use. In sum, the forerunners of the auto possessed intrinsic differences that prohibited further development. The true revolution in automobility would first require a new source of energy that could fulfill the flexible nature needed to support powered human mobility.

HITTING THE ROAD POWERED
BY PETROLEUM

Commodities such as petroleum are culturally constructed: a market must first place a value on them before they are worthwhile. In the earliest years of petroleum, it was refined into kerosene, an illuminant to replace whale oil in lamps. After 1900, when electricity became the source of most lighting, petroleum's greatest value derived from transportation, mainly the automobile. Following the various experiments with other power sources for carriages mentioned above (including steam and electricity), gasoline-driven automobiles were first developed in Europe. A practical gas engine was designed and built by Étienne Lenoir of France in 1860. It ran on illuminating gas. In 1862 he built a vehicle powered by one of his engines. Siegfried Marcus of Austria built several four-wheeled gasoline-powered vehicles. By 1876 Nikolaus Otto, a German, was perfecting his four-stroke cycle engine. Two other Germans, Karl Benz and Gottlieb Daimler, built gasoline cars in 1885.

By the early 1900s many inventors in the United States were also developing new models. In 1893 J. Frank and Charles E. Duryea produced the first successful gasoline-powered automobile in the United States. They began commercial production of the Duryea car in 1896—the same year in which Henry Ford operated his first successful automobile in Detroit. The first automobile salesroom was opened in New York City in 1899 by Percy Owen, and the first automobile show took place in 1900, clearly foreshadowing what would become a dominant technology to define the twentieth century.

From the developments of independent inventors, mass production in the automobile industry was introduced in 1901 by Ransom E. Olds, a pioneer experimenter since 1886. His company manufactured more than four hundred of the now historic curved-dash Oldsmobiles in that first year. Each car sold for only $650. Henry M. Leland and Henry Ford further developed mass production methods during the early 1900s. It remained unclear, though, who actually owned this evolving technology.

In 1879 George B. Selden, an American attorney, had applied for a patent that covered the general features of a gasoline-powered automobile. He received his patent in 1895. In 1903 the Association of Licensed Vehicle Manufacturers was formed by companies that recognized the Selden patent. They agreed to pay Selden a royalty on each car built. Henry Ford refused to join this association. He sued to break Selden's hold on the industry. After extensive litigation, Ford won. In 1911 a District Court of Appeals held that Selden's patent applied only to a two-stroke cycle engine. Other engines were free for the use of other manufacturers. This decision led to a cross-licensing agreement among most of the American manufacturers, which would be administered by the Automobile Manufacturers Association. Under this agreement, the Ford Motor Company organized in 1903, the General Motors Corporation in 1908, and the Chrysler Corporation in 1925.

Using mass production, the first Model T Ford was made in 1908 and more than fifteen million would be sold in the next twenty years. The Model T, nicknamed the "flivver" and the "tin lizzie," was probably more responsible for the development of large-scale motoring than was any other car in automotive history. For power, Ford perfected the internal combustion engine (ICE) that burned gasoline, with which these other inventors experimented as well. During World War I the manufacture of automobiles for civilian uses was virtually halted as the industry was mobilized to produce vehicles, motors, and other war matériel for the armed forces. The automobile assumed a significant new role in the American way of life immediately after the war.[42] No longer an extravagant novelty, the motorcar rapidly became a necessity rather than a luxury for many American families. By the early 1920s most of the basic mechanical problems of automotive engineering had been largely solved. Manufacturers then concentrated their design efforts on making motorcars safer, more stylish, and more comfortable, while also finding new ways to entice consumers and to create the stability and confidence that would encourage car ownership.

In 1929 about 90 percent were drawn from a very few original models; however, diversification of the motoring fleet was underway! By the mid-1920s Henry Ford had decided to abandon the three-pedaled Model T and replace it with the Model A, which was to be equipped with a conventional gearshift. The last Model T was produced in May 1927, and the first Model A rolled off the assembly line in October 1927. An enthusiastic public was soon buying thousands. In 1928 the Chrysler Corporation announced the production of its answer to the Model A—a new low-cost automobile called the Plymouth. The market for moving humans through the landscape appeared to have been won by ICE; however, its true revolution of human patterns of movement would require significant related adaptations.

CURRENT: The United States Models a Society of Drivers

With a foundation of inexpensive power solidified through bountiful supplies of gasoline, auto companies found ways to influence the environment in which the vehicles would be operated. Inconvenience from a lack of roads and infrastructure as well as a reliance on transportation technologies such as trolleys precluded Americans from immediately accepting the new "horseless carriage." The manufacturing and marketing efforts of Ford and others changed this attitude by 1913, when there was one motor vehicle to every eight Americans. Mass production made sure that by the 1920s the car had become no longer a luxury but a necessity of American middle-class life. The landscape, however, had been designed around other modes of transport—including an urban scene dependent on foot travel. Cars enabled an independence never before possible, if they were supported with the necessary service structure. Massive architectural shifts were necessary to make way for the auto. Most important, of course, were the roads on which the autos would travel.

In North America, the first roads were constructed by the Spanish along existing Indian trails into New Mexico and California. As colonists poured into the United States from England, wagon trails were carved along the lines of the Indian footpaths. The first major road system in the United States began to take shape in the late eighteenth century, following the American Revolution, when stagecoaches were in general use and there was an increasing demand for surfaced, all-weather roads. In 1806 Congress authorized construction of the Cumberland, or National, Road, which ran from Cumberland, Maryland, to Vandalia, Illinois. The Cumberland Road opened up the American West.

In the late 1800s, many of the roads in the eastern United States were turnpikes surfaced with tree trunks that were laid across the width of the road to make a so-called corduroy road. Other roads were plank roads, paved with split logs. America's early highway designs were largely the work of Europeans. Two Scottish engineers, John Loudon McAdam and Thomas Telford, pioneered the use of pavements built of broken stone carefully placed in layers and well compacted.

Although the motorcar was the quintessential private instrument, its owners had to operate it over public spaces. Who would pay for these public thoroughfares? After a period of acclimation, Americans viewed highway building as a form of social and economic therapy. They justified public financing for such projects on the theory that roadway improvements would pay for themselves by increasing property-tax revenues along the route. At this time, asphalt, macadam, and concrete were each used on different roadways.

By the 1920s, the congested streets of urban areas pressed road-building into other areas. Most urban regions soon proposed express streets without stoplights or intersections. These aesthetically conceived roadways, normally following the natural topography of the land, soon took the name *parkways*. Long Island and Westchester County, New York, used parkways with bridges and tunnels to separate it from local cross traffic. The Bronx River Parkway (1906), for instance, follows a river park and forest; it also is the first roadway to be declared a national historic site. In addition to pleasure driving, such roads stimulated automobile commuting.[43]

Powerful interests came to clearly see that the automobile's development was in the best social and economic interest of the United States. The Federal Road Act of 1916 offered funds to states that organized highway departments, designating two hundred thousand miles of road as primary and thus eligible for federal funds. More importantly, ensuing legislation also created a Bureau of Public Roads to plan a highway network to connect all cities of fifty thousand or more inhabitants. Some states adopted gasoline taxes to help finance the new roads. By 1925 the value of highway construction projects exceeded $1 billion. Expansion continued through the Great Depression, with road-building becoming integral to city and town development.

Robert Moses of New York defined this new role as road-builder and social planner. Through his work in the Greater New York City area from 1930 to 1960, Moses created a model for a metropolis that included and even emphasized the automobile as opposed to mass transportation. This was a dramatic change in the motivation of design. Historian Kenneth Jackson writes: "In their headlong search for modernity through mobility, American urbanites made a decision to destroy the living environments of nineteenth-century neighborhoods by converting their gathering places into traffic jams, their playgrounds into motorways, and their shopping places into elongated parking lots."[44]

Beginning in the 1920s, legislation created a Bureau of Public Roads to plan a highway network that connected all cities of fifty thousand or more inhabitants in what one historian calls the "largest construction feat of human history."[45] Some states adopted gasoline taxes to help finance the new roads. These developments were supplemented in the 1950s when President Dwight D. Eisenhower included a national system of roads in his preparedness plans for nuclear attack. This development cleared the way for the Interstate Highway Act to build a national system of roads unrivaled by any nation.

When Eisenhower became president, he worked with automobile manufacturers and others to devise a 1956 plan to connect America's future to the automobile. The interstate highway system was the most expensive public works project in human history. The public rationale for this hefty project revolved around fear of nuclear war: such roadways would assist in exiting urban centers in the event of such a calamity. The emphasis, however, was clearly economic expansion. At the cost of many older urban neighborhoods—often occupied by minority groups—the huge wave of concrete was unrolled that linked all the major cities of the nation together.

With the national future clearly tied to cars, American planners began perfecting ways of further integrating the car into American domestic life.

Initially, these tactics were quite literal. In the early twentieth century, many homes of wealthy Americans soon required the ability to store vehicles. Most often these homes had carriage houses or stables that could be converted. Soon, of course, architects devised an appendage to the home and gave it the French name *garage*. From this early point, housing in the United States closely followed the integration of the auto and roads into American life.

In the United States, roads initiated related social trends that added to Americans' reliance on petroleum and wound their lives more tightly to crude than possibly any other society. Most important, between 1945 and 1954, nine million people moved to suburbs. The majority of the suburbs were connected to urban access by only the automobile. Between 1950 and 1976, central city population grew by ten million while suburban growth was eighty-five million. Housing developments and the shopping/strip mall culture that accompanied decentralization of the population made the automobile a virtual necessity. Shopping malls, suburbs, and fast-food restaurants became the American norm through the end of the twentieth century, making American reliance on petroleum complete.[46]

Upper- and middle-class Americans had begun moving to suburban areas in the late 1800s. The first suburban developments, such as Llewellyn Park, New Jersey (1856), followed train lines or the corridors of other early mass transit. The automobile allowed access to vast areas between and beyond these corridors. Suddenly, the suburban hinterland around every city compounded. As early as 1940, about thirteen million people lived in communities beyond the reach of public transportation.[47] Due to these changes, suburbs could be planned for less wealthy Americans. Modeled after the original Gustav Stickley homes, or similar designs from *Ladies' Home Journal* and other popular magazines, middle-class suburbs appealed to working and middle-class Americans. The bungalow became one of the most popular designs in the nation. The construction halt of the Great Depression set the stage for more recent ideas and designs, including the ranch house.[48]

Planners used home styles such as these to develop one site after another with the automobile linking each one to the outside world. The ticky-tacky world of Levittown (the first of which was constructed in 1947) involved a complete dependence on automobile travel. This shift to suburban living became a hallmark of the late twentieth century, with over half of the nation residing in suburbs by the 1990s. The planning system that supported this residential world, however, involved much more than roads. The services necessary to support outlying, suburban communities also needed to be integrated by planners.

Instead of the Main Street prototype, the auto suburbs demanded a new form. Initially, planners such as Jesse Clyde Nichols devised shopping areas such as Kansas City's Country Club District that appeared a hybrid of previous forms. Soon, however, the "strip" had evolved as the commercial corridor of the future. These sites quickly became part of suburban development, in order to provide basic services close to home. A shopper rarely arrived without an automobile; therefore, the car needed to be part of the design program. The most obvious architectural development for speed was signage: integrated into the overall site plan would be towering neon aberrations that identified services. Also, parking lots and drive-through windows suggest the integral role of transportation in this new commerce.

CONCLUSION: CONSPICUOUS CONSUMPTION FOR A GLOBAL SCALE

The new way of life that was built atop massive new supplies of energy changed every aspect of human life, particularly in capitalist nations such as the United States. The matches that enabled ubiquitous illumination and had been so revolutionary in the mid-1800s, now seemed an icon of a distant, primitive era. Historian Jones writes: "Whereas total American energy consumption grew two and a half times [between 1900 and 1930], the use of electricity increased more than twenty-fold."[49] Factories and industries had nearly universal access to electricity and were often charged the lowest rates by utilities. With 20 percent of the American population, mid-Atlantic industries consumed more than 30 percent of the total electricity by 1927. In the first three decades of the twentieth century, energy consumption increased 250 percent, more than twice the rate of population growth. Converting their homes to electricity, Americans between 1922 and 1927 purchased more than ninety-five thousand irons, fifty-four thousand vacuum cleaners, twenty-one thousand washing machines, two thousand sewing machines, and five hundred refrigerators. By 1935, roughly a third of all Americans had a refrigerator. "Electrification furthered the gap between America and the rest of the world," writes Jones. "In 1929, United States electrical production exceeded the combined total of all other countries." These energy changes were sufficient to transform American development, making it the model modern nation.[50] In the United States, mineral energy resources were now not just a luxury but a necessity.

For Americans, these advancements were tied to ideas of prosperity and patriotism, and consumers did not hide their passion for making human life more automated and mechanical. These ideals were presented to the world in venues such as the 1939 World's Fair in New York City. With the United States not yet involved in World War II, Americans used the opportunity to escape the present and wax utopian. Although these dreams took many forms, they were synched together by an invisible hand—more specifically, by a basic assumption that was hidden from each of the scenes: bountiful supplies of cheap energy. Novelist E. L. Doctorow recalled the General Motors Building's "Futurama" exhibit in this fashion:

> In front of us a whole world lit up . . . [and] . . . demonstrated how everything was planned, people lived in these modern streamlined curvilinear buildings, each of them accommodating the population of a small town and holding all the things, schools, food stores, laundries, movies, and so on, that they might need, and they wouldn't even have to go outside.[51]

The exhibition helped to construct a three-dimensional snapshot of a lifestyle entirely different from that of the past. Although unattainable for many human societies, developed nations formed aspirations based on the merging of science-fiction-like visions with the modernist ideas of intellects such as Lewis Mumford. Modernism was no longer an artistic genre restricted to the few artists and designers in European salons; now, modernist design and "the new" were the stuff of the emerging middle class, particularly in the United States. Consumer expectations—such as those shaped by the scene of Futurama—became a primary engine behind these radical shifts in patterns of human living. In the United States, the transformation of living patterns after World War II was so dramatically rapid that historians have given it a name: mass consumption. Historians such as Lizabeth Cohen have demonstrated that this growth in consumption and in the middle class that carried it out was fed by the policies and politics of the Cold War, the ideological conflict that followed World War II. Unseen in such a model of society were the feedback loops of the implications caused by consumption on such a massive scale. In fact, often one could not even discern the primary force behind such ideal societies: cheap petroleum.

Led by Americans, humans invited petroleum into nearly every aspect of their lives during the late twentieth century. Its affordability during most of this era made reliance seem sensible. However, the status of petroleum never changed: even when its supply made it cheap enough to use it in mundane activities such as manufacturing toothpaste tubes, oil was a finite resource—it would run out. By ignoring this reality, though, the American standard of living became the envy of the world and, therefore, expanded the implications of the ecology of oil as other nations sought to model Americans' behavior. Although the results of this new ecology of crude would prove terribly damaging, the post–World War II era of mass consumption marked the blissful, expansionist time of no questions. As the United States used ideology to fight the Cold War, this consumptive society became a great tool for nations that yearned to have a similar standard of living. In short, the view of life on the developed side of the gap drove less-developed nations toward such a goal for themselves.

NOTES

1. Jeremy Zallen, *American Lucifers* (Chapel Hill: University of North Carolina Press, 2019), p. 158.

2. Zallen, p. 171.

3. Zallen, p. 181.

4. Gale Christianson, *Greenhouse* (New York: Walker and Co., 1999), pp. 77–80.

5. Thomas Hughes, *American Genesis* (Chicago: University of Chicago Press, 2004), pp. 184–185.

6. Robert B. Marks, *The Origins of the Modern World: A Global and Ecological Narrative* (Lanham, MD: Rowman & Littlefield, 2002), p. 138.

7. See William Cronon, *Nature's Metropolis* (New York: Norton, 1991).

8. Cronon, p. 120.

9. Cronon, p. 126.

10. Alfred Crosby, *Children of the Sun* (New York: Norton, 2006), p. 112.

11. David F. Burg, *Chicago's White City of 1893* (Lexington: University of Kentucky Press, 1976), p. 204.

12. Hughes, p. 8.

13. Hughes, p. 9.

14. Stephen Kern, *The Culture of Time and Space* (Cambridge: Harvard University Press, 2003), p. 210.

15. See Brian C. Black, *Crude Reality* (New York: Rowman & Littlefield, 2020).

16. Daniel Yergin, *The Prize* (New York: Free Press, 2008), p. 56.

17. Steve LeVine, *The Oil and the Glory* (New York: Random House, 2007), p. 68.

18. Benjamin Labaree et al., *America and the Sea* (Mystic, CT: Mystic Seaport Museum 1998), pp. 390–391.

19. Richard Rhodes, *Energy* (New York: Simon & Schuster, 2019), p. 168.

20. Rhodes, pp. 174–176.

21. Hughes, pp. 226–230.

22. Hughes, pp. 234–240.

23. Hughes, pp. 39–40.

24. David Nye, *Electrifying America* (Cambridge, MA: MIT University Press, 1992), pp. 138–142.

25. David Nye, *American Technological Sublime* (Cambridge, MA: MIT University Press, 1996), pp. 144–148.

26. John R. McNeill, *Something New Under the Sun: An Environmental History of the Twentieth-Century World* (New York: Norton, 2001), pp. 174–175.

27. Christian Anton Smedshaug, *Feeding the World in the 21st Century* (London: Anthem Press, 2010), 49–50.

28. Smedshaug, pp. 180–181.

29. See, for instance, Richard Tucker, *Insatiable Appetite* (New York: Rowman & Littlefield, 2007).

30. Michael Pollan, *Omnivore's Dilemma* (New York: Penguin, 2006), p. 44.

31. Pollan, p. 45.

32. Marks, pp. 165–166.

33. Lewis Mumford, *Technics and Civilization* (Chicago: University of Chicago Press, 2010), p. 322.

34. Leland Roth, *Concise History of American Architecture* (New York: Harper & Row, 1980), pp. 161–162.

35. Adrian Smith and Judith Dupre, *Skyscrapers* (New York: Black Dog & Leventhal, 2013), pp. 242–243.

36. Alan Trachtenberg, *Incorporation of America* (New York: Hill & Wang, 2007), p. 35.

37. John Kasson, *Amusing the Million* (New York: Hill & Wang, 1978), p. 32.

38. Judith Adams, *American Amusement Park Industry* (New York: Twayne Publishers, 1991), p. 46.

39. Brian Black, *Petrolia* (Baltimore: Johns Hopkins University Press, 2000).

40. Black, *Petrolia*.

41. Erik Eckermann, *World History of the Automobile* (New York: Society of Automotive Engineers, 2001), pp. 9–11.

42. Black, *Crude Reality*.

43. John Stilgoe, *Common Landscape of America* (New Haven, CT: Yale University Press, 1983), p. 23.

44. Kenneth T. Jackson, *Crabgrass Frontier* (New York: Oxford University Press 1985), p. 168.

45. Tom Lewis, *Divided Highways* (New York: Penguin, 1997).

46. See Jackson, *Crabgrass Frontier*.

47. See Jackson, *Crabgrass Frontier*, p. 140.

48. See Jackson, *Crabgrass Frontier*, p. 140.

49. Christopher Jones, Routes of Power (Cambridge: Harvard University Press, 2016), pp. 217–219.

50. Jones, pp. 221, 225.

51. E. L. Doctorow, *World's Fair* (New York: Random House, 1985), pp. 252–253.

III

ENERGY BROADENS THE GAP (1900–2000)

Throughout the world, natural seeps of gas and oil became important signals of future supplies. LIBRARY OF CONGRESS PRINTS AND PHOTOGRAPHS DIVISION

In describing the global human of the 1910s and '20s as going through a "Great Departure," historian Marks writes that World War I "shook the imperialist order to its foundation and had major consequences for the shape of the 20th c. world."[1] Energy served as a catalyst and a tool for the growing disparity between haves and have-nots. In the Middle East, petroleum reserves emerged as a rationale for lingering colonialism as well as emerging independence. Natural seeps of gas or oil in the African desert provided the setting for this image from the 1930s. Symbolically, the suited man straddles the gap between developed and less-developed nations embodied in the potential of the energy source held beneath his feet.

NOTE

1. Robert Marks, *The Origin of the Modern World* (Lanham, MD: Rowman & Littlefield, 2019), pp. 161–162.

5

Energy and National Security

GROUND: COAL STATION, BORNEO, PACIFIC OCEAN

The geography of the Age of Sail had been fairly straightforward; coal-powered shipping, however, required new, more complex infrastructure. By relying on a non-renewable prime mover, shipping companies needed to organize their routes around a system of recharging ports, known as coal stations. While coal could be brought in and stored in existing ports, a very few other sites became popular because they offered mines to produce their coal. As Britain and the United States opened up Pacific trade in the mid-1800s, the coal of Borneo became a strategic advantage of the greatest navies and traders in the world.

Knowledge of Borneo's coal resource came about entirely by chance through the early trade contacts of the American merchant house Canton, Olyphant and Company. George Tradescant Lay, who served as Far Eastern representative for the company and its ancillary Bible society, visited Borneo in 1837 and during a meeting at the sultan's palace was presented with a piece of local coal. Intrigued, he searched the area and excavated a small seam of coal. Lay wrote of his discovery two years later to the great interest of British and American governments. In 1842, the British governor of Bengal appointed agents to test Borneo's coal for use in steam shipping. As historian Peter Shulman writes: "No other part of the world offered the British greater promise of commercial gain and yet presented greater obstacles to the flow of goods and information. . . . With communication would come trade, and with trade, wealth and power."[1]

Although the British head start was significant, American emphasis on steam shipping by 1850 made access to Asia a realistic proposition. Commodore Matthew Perry joined these initiatives by emphasizing that mail steamers met naval specifications so that they could be used for such purposes in an emergency. He famously led an 1852 excursion through the Pacific in order to denote the necessary coaling sites with a clear emphasis on Borneo. While Americans at home debated the connection between coal and foreign policy, Perry navigated in the realm of realism. He scoured the region for possible coaling locations.

As an emerging center for trade with the Asian world, Pacific ports such as Borneo became critical for their ability to recharge the energy of shippers. By the late 1800s, when coal-powered navies had become a prime measure of strategic global power, these same coaling stations took on even larger importance.

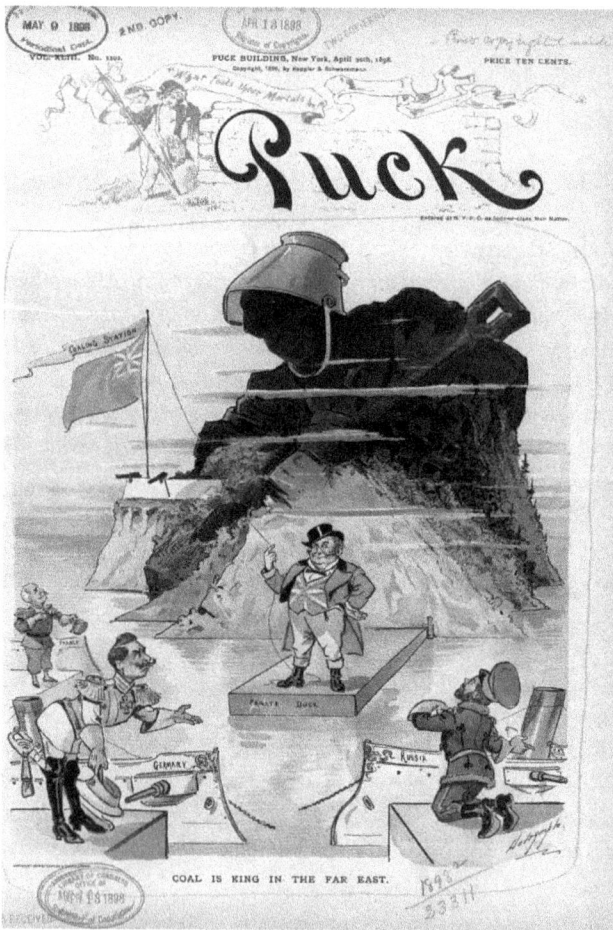

Coal remained an essential portion of national defense through the twentieth century. LIBRARY OF CONGRESS PRINTS AND PHOTOGRAPHS DIVISION

From sailing ships to bombards, the management of energy supplies had proven to be an important tool for national security and expansion. With the twentieth century's emphasis on fossil fuels, though, new technologies based on ensuring a foundation of massive supplies of energy became tools of national power. Particularly through the experience of warfare up to 1950, strategic thinkers devised methods of applying emerging technologies that helped to define or establish political and economic power on the global stage.

Historian Marks joins others in describing the global human of the 1910s and 1920s as going through a "Great Departure," and he writes that World War I "shook the imperialist order to its foundation and had major consequences for the shape of the 20th c. world." The combination of both rapid industrial and population growth in the twentieth century redefined humans' relationship to their environment and clearly separated them from the rhythms and constraints of the biological old regime. During this era, foundational transitions in energy consumption are compounded by applications that multiply fossil fuels' importance, including the creation of synthetic fertilizer for both agriculture and military use as explosives and, of course, powering personal transportation. In these examples and many others, the expansive supply and expansive applications of ubiquitous energy stirred intrinsic shifts in the human condition.

In the case of the general expansion of energy use, we see new technologies of the era provide a shrinking of time and space as well as an intensification of human impacts and potentialities. In *Something New Under the Sun*, John McNeill writes that worldwide energy harvest increased by about five times in the nineteenth century and sixteen times in the twentieth. The Great War (or World War I) sits at a defining precipice of this shift in our species. From deforestation and mining to the expansion of automobile use and the internal combustion engine, we find that the World War I era played a formative role in remaking the human condition. Indeed, as a divider between nineteenth- and twentieth-century ways of life, the Great War demarks an entirely new scale and scope to energy use and development. As scholars reorient the human story to better reflect these modal distinctions in the way that people live, World War I also becomes the gateway to the "Anthropocene."

The energy transition of the World War I era only succeeded because of the advancement of additional technologies that enhanced the impact of new sources of power. Historian Christopher Jones discusses the multiplying effects of such related or ancillary technologies as "intensification" that took the shift to mineral-based sources begun in the nineteenth century and formalized and broadened their applications.[2] With such new energy systems in mind, the summary impact of the Great War on energy management emerges as many layered and complex but altogether transformative. The stage for this transition, however, was set just prior to the conflict and then fully played out during World

War II. The era from 1914 to 1945, thereby, becomes much more than a simple transition between sources of power; instead, this era escalates the primacy of energy in human life to entirely new scales. As the capabilities of the "Great Acceleration" became evident, so, too, did its potential impacts.

CONDUIT: Oil Derrick, Spindletop, Texas, 1901

Without warning, the level plains of eastern Texas near Beaumont abruptly give way to a lone, rounded hill before returning to flatness. Geologists call these abrupt rises in the land domes, because hollow caverns lie beneath. Over time, layers of rock rise to a common apex and create a spacious reservoir underneath. Often, salt forms in these empty, geological bubbles creating a salt dome. Over millions of years, water or other material might fill the reservoir. At least, that was Pattillo Higgins's idea in eastern Texas during the 1890s.

Higgins and others imagined such caverns as natural treasure houses. Few listeners shared their sentiment, however. With plenty of oil to be drilled in areas where it was known to occur, most oilmen scoffed at Higgins. Their derision only made him more committed to the search for Texas oil. In particular, Higgins's intrigue grew with one dome-shaped hill in southeast Texas. Known as Spindletop, this salt dome—with Higgins's help—would change human existence.

Although the search for oil moved into a variety of stretches of the globe, Texas had not yet been identified as an oil producer. In the United States, well-known oil country lay in the east, particularly western Pennsylvania. In addition, Standard Oil and other companies explored far-off regions such as Baku. By the 1890s, petroleum-derived kerosene had become the world's most popular fuel for lighting. Thomas Edison's experiments with electric lighting placed petroleum's future in doubt; however, petroleum still stimulated boom wherever it was found. But in Texas? Every geologist who inspected the "Big Hill" at Spindletop told Higgins that he was a fool.

Thanks to the rule of capture, the independent speculators, such as Higgins, were a critical part of petroleum from its beginning in Pennsylvania. Before corporate monoliths, petroleum represented one of the great American examples that one man could make an incredible difference—and fortune—if he was willing to work hard and take calculated risks.

Today, these companies use science (particularly seismic locating technology) to locate underground supplies and diminish much of the unknown qualities of crude. They cost-benefit the size of the reserve with

Spindletop, Texas, in 1901 introduced the world to massive supplies of crude oil. Now we had to find something to do with it! LIBRARY OF CONGRESS PRINTS AND PHOTOGRAPHS DIVISION

the expenditures required to get it out and only pursue those that will turn a profit. And yet all the glamour has not dissipated. At the ground level, there is a clear continuity from Drake to Higgins and even to the petroleum corporations of the twenty-first century. Even though it pales in comparison to the corporate culture today, by 1900 petroleum speculation was serious business because petroleum was seriously valuable. The growth of the infrastructure used to gather, process, and distribute crude in the early twentieth century formed one of the most dominant global corporate organizations of the entire century. Efficiency and effectiveness created corporate profitability by the early 1900s and this helped to make petroleum one of the cheapest sources of energy in human history.[3]

Harvesting oil in 1901 very much resembled efforts used in the Pennsylvania oil fields of the 1860s and, in fact, those of today. Surface seepages or other evidence led speculators to choose a place to drill through earth's crust, similar to methods for drilling a well for water. Once a significant enough flow was achieved, a derrick was set atop the well that would pump the oil to the surface and regulate the flow of oil. Oil fields became very recognizable by the slow but constant swing of the derrick arms that were attached to towers and normally surrounded by tanks that could be used for separating the oil from the water that the derrick brought upward. Normally, striking an oil well was a fairly docile industrial activity; however, particularly as speculators reached deeper into Earth, the oil reservoir was often full of pressure—normally due to the presence of natural gas. In such a situation, the drill punctured the pressurized reservoir and created a gusher in which the oil sprayed out of the well and into the sky like a geyser. Dangerous, wasteful, and sloppy, gushers were frowned upon by those trying to manage and systematize oil production.

Spindletop came in as the wildest gusher that had yet been seen. At Spindletop on January 10, mud began bubbling from the drill hole. The four to five workers at the site struggled to make sense of the mud just as they heard a sound like a cannon. A roar followed, and suddenly oil spurted out of the hole with rock, sand, and portions of the drilling rig flying through the air. The Lucas geyser, found at a depth of 1,139 feet, blew a stream of oil over a hundred feet high until it was capped nine days later. During this period, the well flowed an estimated one hundred thousand barrels a day—well beyond any flows previously witnessed. After a few months over two hundred wells had been sunk on the Big Hill, and by the end of 1901 an estimated $235 million had been invested in oil in Texas.[4] The production in Spindletop and East Texas oil altered the scale of the entire industry after 1901. The energy transition of this era was largely driven by abundant—even overwhelming—supply and not by one specific technical innovation. At this moment of overwhelming supply, though, petroleum had very few applications.

BRITAIN CONNECTS COAL AND OIL WITH NATIONAL SECURITY

Logistics, such as the coal station in Borneo and elsewhere, marked a by-product of the transition from wind to coal power in shipping. The source of sailing routes had partly derived from the patterns and cells in which wind circulated in great circles. While shifting to coal did not immediately obviate the use of the popular route in the Pacific, for instance, Shulman writes, "fuel was expensive, and as both commercial and naval steamships grew in size and strength, minimizing a ship's steaming distance offered a way to stretch resources beyond existing limits of engine design and available coaling depots."[5] On board, fueling coal burners involved constant labor, and storing the bulky fuel required great amounts of stowage. Despite these limitations and the obvious appearance of coal as only a temporary solution to fueling ships, in the later decades of the nineteenth century a transition occurred on trade and passenger lines as well as national navies. Clearly, though, the transition from sail to coal seemed only temporary.

Emerging from petroleum's initial commercial development in Pennsylvania during the 1860s, subsequent decades brought international speculation and increased supply (from locations such as Texas) but few new uses. A unique petroleum culture took shape around this expanded production and availability. Such a moment, partly created by technical and corporate innovation and partly by new roles for the nation-state and consumer ideas of individual autonomy, ushered in a revolutionary assortment of new ideas for how crude might be applied to human living.

Although the popularity of kerosene had jettisoned petroleum's value in the nineteenth century, illumination appeared a fleeting application of crude by the late 1800s. Tinkerers and scientists in Europe and the United States began extensive application of electricity for a variety of purposes. Most important, Edison and other inventors and system-builders joined to make electricity (created from coal or other means) the most likely illuminant of the future. By the 1890s, therefore, the primary use of petroleum was becoming lubrication and manufacturing. Ironically, at this same moment, international efforts to develop and acquire petroleum intensified. In the first decade of the twentieth century, therefore, resourceful political leaders sought ways to make petroleum useful and, even, integral in ways that extended beyond lubrication. In both the United States and Great Britain, the energy transition was openly maneuvered and manipulated by the political and military establishment. As early as 1910, petroleum's growing abundance had helped it to emerge as a strategic tool for ensuring global power. The first application for petroleum in this regard was ensuring naval supremacy for each of these global powers.

Politically, the British effort at naval conversion was eventually led by young Winston Churchill, who began as a member of Parliament and by 1910 had become the president of the Board of Trade. Although he did not begin on the side of naval expansion, the early 1910s brought Churchill a clear education on the advantages of oil (speed capabilities, flexibility of storage and supply, permitted refueling at sea, etc.). He later wrote:

> As a coal ship used up her coal, increasingly large numbers of men had to be taken, if necessary from the guns, to shovel the coal from remote and inconvenient bunkers to bunkers nearer to the furnaces or to the furnaces themselves, thus weakening the fighting efficiency of the ship perhaps at the most critical moment in the battle. . . . The use of oil made it possible in every type of vessel to have more gun-power and more speed for less size or less cost.[6]

By 1912, the policies had been put in place and, as Churchill recorded, in the world's greatest navy "the supreme ships of the Navy, on which our life depended, were fed by oil and could only be fed by oil." Churchill and Britain's military strategists emphasized the great benefits for their naval superiority; however, their decision also marked a defining moment in a new era of the culture of petroleum. By association, committing their fleet to petroleum meant that a consistent and reliable supply of petroleum had just become one of the most important commodities on Earth—nations' security depended on it. Also, by association, any nation wishing to compete with them had to follow suit.

By declaring this new energy era, Britain forced any competing nations to also consider oil with this new logic. At the highest levels, U.S. leaders debated the implications of converting their military, particularly the navy, to petroleum. Their conversations had begun in the late 1800s but took on greater urgency as British reconversion altered global affairs. The United States had one significant strategic advantage in the area of naval conversion: in the early 1900s, American oil fields produced approximately one-third of the world's oil. Indeed, the United States approached all such strategic decisions from the basic realization that it was the only nation in the world that could power its military with petroleum and largely be able to supply it from its own reserves—it possessed energy autonomy, which would later become known as energy independence. Although this was an obvious advantage over other nations, the American situation also required a new type of relationship between business and government. Given such critical importance, petroleum's supply demanded federal oversight or management. In times of an overabundance of supply, this control was often referred to as conservation.

In the end, though, it was Churchill who seems to have most clearly formed the necessary new vision of the twentieth-century world when he proclaimed to the House of Commons on June 17, 1914:

> This afternoon we have to deal, not with the policy of building oil-driven ships or of using oil as an ancillary fuel in coal-driven ships. . . . Look out upon the wide expanse of the oil regions of the world. Two gigantic corporations—one in either hemisphere—stand out predominantly. In the New World there is the Standard Oil. . . . In the Old World the great combination of the Shell and the Royal Dutch.
>
> For many years, it has been the policy of the Foreign Office, the Admiralty, and the Indian Government to preserve the independent British oil interests of the Persian oil-field, to help that field to develop as well as we could and, above all, to prevent it being swallowed by [others].
>
> [Over]. . . the last two or three years, in consequence of these new uses which have been found for this oil. . . . There is a world shortage of an article which the world has only lately begun to see is required for certain special purposes. That is the reason why prices have gone up, and not because evilly-disposed gentlemen of the Hebraic persuasion.[7]

Therefore, on the eve of World War I, the status of crude had been altered dramatically and its new importance would be pressed on a global stage almost immediately.

CURRENT: Price, Stratigraphy, and Global Competition for Oil

In Great Britain, when Churchill committed the Royal Navy to petroleum in 1913, he forever compromised the nation's energy autonomy: Britain had neither domestic sources of oil nor existing supplies in its colonies. Anglo-Persian/BP, with its access to oil in Central Asia (particularly Persia, the future Iran), quickly became the most sensible option to ensure Britain's energy future. Large capital expenditures, such as pipeline construction, had left Anglo-Persian/BP in deep debt and near bankruptcy by 1914. To convince Parliament to help the company, Churchill argued: "If we cannot get oil, we cannot get corn, we cannot get cotton and we cannot get a thousand and one commodities necessary for the preservation of the economic energies of Great Britain."[8] Parliament approved his plan to purchase a 51 percent stake in BP for £2.2 million in June 1914. Maintaining and developing oil supplies soon became a critical portion of the British colonial efforts.

Global use of petroleum grew by 50 percent during World War I, which exacerbated the difficulty of managing the global supply. These difficulties grew more acute in 1919 when one of the world's significant producing regions, Russia, destabilized. The United States, however, still dominated oil production in 1919, producing one million barrels or approximately 70 percent of the global output. But the culture of oil was changing.

One reason for what experts referred to as an "oil glut" was an increased understanding of oil geophysics, which began with the U.S. Geological Survey in 1908 and industrial publications such as the *Oil & Gas Journal*, which began publication in 1902. Particularly in the United States, the changing culture of petroleum in the 1920s moved serious science into the petroleum industry. Primarily, new understanding allowed regulation and control to begin to reign in the rule of capture that had held that oil fields were fair game to any wildcatter. Perceiving the unity of oil fields—meaning the connectivity of underground wells—was substantiated by theories including the anticlinal theory (which demonstrated the underground occurrence of natural gas, oil, and water and tied them to surface features such as domes) and subsurface mapping, which grew into stratigraphy after World War I. Using stratigraphic features of geology, drilling exploratory fields became much less a leap of faith and much more a technical, scientific certainty administered by corporate managers. This also contributed to the development of secondary recovery, which involved flushing out existing wells with natural gas to acquire reserves previously inaccessible.[9] Big Oil was able to acquire the oil that it pursued at a higher rate than ever before.

In addition, chemical science applied to refining led to the breakthrough of thermal cracking that was introduced in 1913 by William Burton and allowed crude to be systematically separated into a variety of products. Most importantly, gasoline acquired from each barrel of crude grew from 15 percent of American production in 1900 to 39 percent in 1929.[10] In the emerging era of Big Oil, corporate entities such as Shell and BP integrated each of these new sciences to gain a mixed profile while others specialized—for instance, Texaco, Chevron, and Gulf specialized on locating and harvesting crude while Exxon focused on refining. As gasoline emerged as a primary output, each corporation also gained a public face in which it interacted with consumers through gas stations. Overall, writes industry veteran Leonardo Maugeri, "This transformation of the industry into its modern shape involved a vast process of mergers and acquisitions, favored by its growing capital intensity. . . . Mergers and acquisitions proved a quicker and more profitable way to achieve integration, scale, and market presence than building them step-by-step."[11] As one energy source, oil was reaching a global level of importance never before seen.

CONFLICT AND INNOVATION IN THE GREAT WAR

The catalyst for many of these transitions in energy use was conflict on a global scale. In determining humans' transportation future, for example, the brief explanation is simple: World War I relied on the use of vehicles, and the electric-powered alternatives that were succeeding in many portions of the consumer market could not fulfill the flexibility required for warfare. During World War I the manufacture of automobiles for civilian uses was virtually halted as the industry was mobilized to produce vehicles, motors, and other war matériel for the armed forces. The role of automobiles for use in the war effort emerged immediately when a fleet of Parisian taxicabs were used to bring troop reinforcements forward during the Battle of the Marne in 1914.[12]

It is estimated that 125,000 Ford Model Ts were used by the Allies on the battlefield of World War I. In addition, truck production was doubled. Even though the American auto and truck industry was required to make other products as well (shells, guns, etc.), increased vehicle needs actually allowed the industry to increase production during the war. Historian David Kirsch notes that in France, Britain, Germany, and later Russia, truck purchasers received up to $1,200 from the government for the purchase of an approved vehicle—which stipulated petroleum, internal-combustion (ICE)-powered vehicles over the electric alternatives. U.S. manufacturers established designs for a standard war truck in 1916–1917 and consequently began exporting vehicles to the front. Kirsch writes: "The dramatic role of motor trucks in the conduct of the Great War reinforced and accelerated the standardization of the commercial peacetime truck. . . . By 1919 electric trucks accounted for less than 1 percent of the total number of commercial vehicles produced in the United States, down from 11 percent in 1909."[13]

Mobility on the U.S. home front was influenced in basic ways by the needs of the war. For instance, in the United States the strain on the nation's railroads fueled the military to emphasize long-distance trucking and to call for the roads that these routes made necessary. In addition, most trucks for the war effort were manufactured in the Midwest and needed to be brought to the eastern seaboard for shipment abroad. It is estimated that during 1917–1918 eighteen thousand ICE-powered trucks made this trip. In the United States, these trips, which were driven be necessity, demonstrated that such vehicles could be used reliably for interstate shipping, in lieu of railroads. Federal funds had begun to develop such roads in 1912, which primarily focused on rural access for the U.S. Postal system. In 1916, the Federal Aid Road Act focused federal funds on roads that would help farmers to get their products out of rural areas with more ease and flexibility.[14]

On each battlefield of the twentieth century, energy—and particularly fossil fuel supply—became an integral portion of every strategic decision. LIBRARY OF CONGRESS PRINTS AND PHOTOGRAPHS DIVISION

Following the war, commercial trucking in the United States became a dramatic example of technological selection—consumers selected the ICE-powered vehicle over the electric alternative. The dominant form of commercial transport within urban areas remained horsepower; however, electric-powered trucks seemed a superior alternative for short-haul delivery systems. After World War I, though, explains Kirsch, standard practices within the industry—including the use of long-haul trucking over railroad—forced the "appropriate sphere of the electric truck [to grow] smaller and smaller."[15] Although proponents of electrics pushed for separate spheres of transportation with separate technologies, business owners could not support hybrid fleets. In making their decision for ICE-powered trucks, businesses accepted a cost-benefit scenario that allowed them to succeed across the board, even if another technology (electric power) made more sense for short-hauling. It was these decisions, fed by cheap fuel prices and government-sponsored infrastructure, that helped to determine the future patterns in human mobility.

During World War I, having a domestic abundance of crude to manage put the United States in a powerful position, no matter what regulatory choices it made. For other developed nations that did not possess petroleum, their growing reliance dictated new patterns of trade and diplomacy. Before the supply was required by the navy, though, it was proven indispensable on the battlefield. The energy transition that stemmed from Britain's commitment to petroleum had obvious influence on battlefield strategy, first in World War I and then in World War II. Less noted by historians, petroleum's importance in matters of national security and diplomacy frames the grander patterns of this era, tying together each war and also the interregnum that separates them.

Although neither conflict grew entirely from disagreements associated with petroleum supply, new systems of negotiation and need had emerged that would eventually be referred to as "geopolitics" to include concepts such as spheres of influence and trade, which were dictated by location as well as specific resources that were needed. By considering some of these larger patterns during 1915–1945, we see petroleum moving to the front of the many concerns that composed the developed nations' ideas of energy security. Conflict was increasingly less focused on disputes of border and more often emphasized important resources such as energy and particularly the fickle supplies of petroleum that occurred in very limited locations.

The European conflict mentioned above, of course, became known as the Great War or, in historical retrospect, World War I. It was a disastrous blend of Old World strategies and new, modern-era technologies. Thirteen million people died and millions more were injured or had their lives decimated. "It was a war," writes Yergin, "that was fought between men and machines. And these machines were powered by oil."[16]

When Britain set out to create its petroleum-powered navy during the spring of 1914, Europe seemed more peaceful than it had for years. Just eleven days after Parliament approved Churchill's bill about petroleum, though, Archduke Franz Ferdinand of Austria was assassinated at Sarajevo. Russia's army mobilized on July 30, and Germany declared war on it on August 1. British hostilities against Germany began three days later on August 4, 1914.

When the war broke out, military strategy was organized around horses and other animal participants. With one horse on the field for every three men, such primitive modes dominated the fighting in this "transitional conflict." Throughout the war, the energy transition took place from horsepower to gas-powered trucks and tanks and, of course, to oil-burning ships and airplanes. Innovations put new technologies into immediate action on the horrific battlefield of World War I. It was the British, for instance, who set out to overcome the stalemate of trench warfare by devising an armored vehicle that was powered by the internal combustion engine. Once again, Churchill is given credit for bringing the project—under its code name "tank"—to reality when other British politicians wished to continue with existing practices. Although the tank was first used in 1916 at the Battle of the Somme, its decisive use arrived in August 1917 when at the Battle of Amiens a squadron of nearly five hundred British tanks broke through the German line. In addition, the British Expeditionary Force that went to France in 1914 was supported by a fleet of 827 motorcars and fifteen motorcycles; by war's end, the British army included fifty-six thousand trucks, twenty-three thousand motorcars, and thirty-four thousand motorcycles.[17] These gas-powered vehicles certainly offered superior flexibility on the battlefield; however, their impact on land-based strategy would not be fully felt due to the continued prevalence of other methods of fighting.

In the air and on the sea, the strategic change was more obvious. By 1915, Britain had built 250 planes. In this era of the Red Baron and others, primitive airplanes often required that the pilot pack his own sidearm and use it for firing at his opponent. More often, though, the flying devices could be used for delivering explosives in episodes of tactical bombing. German pilots applied this new strategy to severe bombing of England with zeppelins and later with aircraft. Over the course of the war, the use of aircraft expanded remarkably: Britain, fifty-five thousand planes; France, sixty-eight thousand; Italy, twenty thousand; United States, fifteen thousand; and Germany, forty-eight thousand.[18] The disagreement over using petroleum at sea helped to exacerbate existing conflict leading up to the war. Ironically, the use of petroleum in ships led to what Yergin calls a "stalemate" in the use of ships during the war, with only one battle (the Battle of Jutland in 1916). However, part of the explanation for this is the great chasm that separated Britain's emerging petroleum-powered shipping fleet from the entirely coal-burning one of Germany. It made little strategic sense for

Germany to confront the British Navy; therefore, it used the tactic of submarine warfare. These early submarines ran primarily as diesel-powered ships on the surface, which were capable of briefly diving for attacks while they ran on battery power.

With these new uses, wartime petroleum supplies became a critical strategic military issue. Royal Dutch/Shell provided the war effort with much of its supply of crude. In addition, Britain expanded even more deeply in the Middle East. In particular, Britain had quickly come to depend on the Abadan refinery site in Persia and when Turkey came into the war in 1915 as a partner with Germany, British soldiers defended it from Turkish invasion. In addition, British soldiers pushed in to take control of Basra and eventually Baghdad. These defensive efforts allowed British fuel to continue to come from the Abadan refinery. Oil production in Persia grew during the war from sixteen hundred barrels per day (bpd) to eighteen thousand bpd. Of course, the growth and stability of the supply grew from Britain's Anglo-Persian corporation, which by the end of the war had purchased the British Petroleum distribution company from the Crown. In order to move the supply where it was needed, the company quickly became a pioneer in the tanker business. By 1917, for just the reasons listed above, oil tankers had become one of the German sub fleet's favorite targets. Late in this year, the loss of tankers had become so extreme that British leaders worried that the war effort would be stymied.

When the Allies took renewed measures to prosecute the war in 1918, petroleum was a weapon on everyone's mind. The Inter-Allied Petroleum Conference was created to pool, coordinate, and control all oil supplies and tanker travel.[19] The U.S. entry into the war made this organization necessary because it had been supplying such a large portion of the Allied effort thus far. As the producer of nearly 70 percent of the world's oil supply, the United States's greatest weapon in the fighting of World War I may have been crude. President Woodrow Wilson appointed the nation's first energy czar whose responsibility was to work in close quarters with leaders of the American companies. These policies began more than a century of close relations between the U.S. government and oil executives—Big Oil. As a result of this cooperative relationship, when domestic prices for crude rose during the war, the czar made an appeal for "gasolineless Sundays" and other voluntary conservation measures.

On the battlefield, the Allies also designed their strategy to disrupt even the limited supply of petroleum. Although Germany was heavily dependent on the Romanian oil fields, the small nation refused to join it in the war effort. Finally, in 1916, Romania declared war against Germany. As a result, German troops advanced on the oil fields and stored reserves. With Rumania's limited ability to rebuff Germany's advances, Britain moved forward on its own approach to the problem: to destroy the Romanian industry so that it could be of no assistance to their opponent. By the end of 1916, British explosives had been used to relegate the entire Romanian industry—fields, reserves, and other apparatus—to waste. The destruction was total; however, Germany took back the Romanian fields and by 1918 had restored approximately 80 percent of the oil supply. In addition, Germany had made significant movement toward acquiring the Baku supply after the Russian Revolution of 1917. More rapidly, though, their own ally Turkey advanced on the valuable resource, suddenly unguarded. By mid-1918, British forces responded to Baku's cries for help and arrived to defend the fields from the Germans—including instructions to destroy them if their defense became untenable.

Yergin writes that the denial of Baku's supply at this juncture proved "a decisive blow for Germany."[20] At the meeting of the Inter-Allied Petroleum Conference immediately after the armistice had been signed, the lead speaker declared: "The Allied cause had floated to victory upon a wave of oil." A later speaker from France offered that just as oil had been the blood of war, now it must "be the blood of the peace."[21] This realization defined most human lives during the coming decades as petroleum became a critical domestic commodity.

More important, though, as a strategic commodity, petroleum would never leave center stage. As Woodrow Wilson led world leaders to think cooperatively of a League of Nations, British forces secured their control over Mesopotamian oil by taking Mosul. In addition, ensuing agreements secured British dominance over the area now known as the Middle East. Their interest fueled further exploration by oil companies and by petroleum geologists. By the 1920s, the findings established a red line spanning the nations reaching from Turkey to Oman that held the largest supply of petroleum on Earth. By 1928, this arrangement took more official form as the "Red Line Agreement," in which Royal Dutch/Shell, Anglo-Persian, an "American Group" (five private companies), and French interests agreed only to work within this region in cooperation with the Turkish Petroleum Company, which was led by Calouste Sarkis Gulbenkian, an Armenian entrepreneur who was responsible for the finding. Members of the group were given a 23.75 percent share in the consortium and asked to subscribe to a self-denying ordinance that prohibited the members from engaging in independent oil development within the designated region.[22] Clearly, a new world order had begun to form around energy supplies.

AMERICA'S GREAT CONVOY, CATALYST FOR ENERGY TRANSITION

On July 7, 1919, a group of U.S. military members dedicated Zero Milestone just south of the White House lawn in Washington, D.C. The next morning, they helped to define the domestic future of the nation that they served. Instead of an exploratory rocket or deep-sea submarine, these explorers set out in forty-two trucks, five passenger cars, and an assortment of motorcycles, ambulances, tank trucks, mobile field kitchens, mobile repair shops, and Signal Corps search-light trucks. During the first three days of driving, they managed just over five miles per hour. This was most troubling because of their goal: to explore the condition of American roads by driving across the country.

Leading this exploratory party was U.S. Army captain Dwight D. Eisenhower. Although he played a critical role in many portions of twentieth-century U.S. history, his passion for roads might have carried the most significant impact on the domestic front. This trek literally and figuratively caught the nation and the young soldier at a crossroads. Returning from World War I, Ike was entertaining the idea of leaving the military and accepting a civilian job. His decision to remain proved pivotal for the nation. By the end of the first half of the century, the new roadscape would remake the nation and the lives of its occupants. For Ike, though, roadways represented not only domestic development but also national security.[23]

This realization is embodied in the person of Eisenhower, who termed the travelers' progress over the first two days "not too good" and as slow "as even the slowest troop train." Ike described the roads they traveled as "average to nonexistent." He continued:

In some places, the heavy trucks broke through the surface of the road and we had to tow them out one by one, with the caterpillar tractor. Some days when we had counted on sixty or seventy or a hundred miles, we could do three or four.[24]

Eisenhower's party completed its frontier trek and arrived in San Francisco on September 6, 1919. Similar to Lewis and Clark and the builders of the transcontinental railroad, their effort proved to be a pivotal national moment. The significance of their one step for mankind, though, culminated with neither a golden spike nor an American flag; instead, they sparked an awareness that initiated a century's worth of reaction from Americans. Of course, the clearest implication that grew from Eisenhower's trek was the importance of roads. Unstated, however, was the symbolic suggestion that matters of transportation and of petroleum now demanded the involvement of the U.S. military as it did in many industrialized nations.

In typically understated language, Ike's recollection misdirects listeners from the dramatic shift that lay below the surface. "The old convoy," he explained, "had started me thinking about good, two lane highways."[25] The emphasis on roads and particularly on Ike's interstate system was transformative for the country; however, Eisenhower was overlooking the fundamental shift in which he participated. The imperative was clear: whether through road-building initiatives or through international diplomacy, the use of petroleum by his nation and others was now a reliance that carried with it implications for national stability and security.

Seen through this lens of human history, petroleum's road to essentialness in human life begins neither in its ability to propel the Model T nor to give form to the burping plastic bowl. The imperative to maintain petroleum supplies begins with its necessity for each nation's defense. Although petroleum use eventually made consumers' lives simpler in numerous ways, its use by the military fell into a different category entirely. If the supply was insufficient, the nation's most basic protections would be compromised. In 1919, Eisenhower and his team thought they were only determining the need for roadways. In fact, they were declaring a political commitment by the United States that would guide diplomacy for decades. And, in fact, thanks to its immense domestic reserves, the United States was late coming to this realization; it was a commitment already being acted upon by nations that lacked essential supplies of crude.

The reliance of the military on petroleum set the tone for humans' twentieth-century commitment to crude. Unlike any other resource, petroleum received its own administrative infrastructure at the highest levels of government once it served as the lifeblood of military infrastructure. Historian David S. Painter writes: "The result was a public-private partnership in oil that achieved U.S. political, strategic, and economic goals, accommodated the desires of the various private interests, conformed to U.S. ideological precepts, and palliated congressional critics."[26] Out of the marriage of security and petroleum, twentieth-century America received a less noticeable but even more critical rationale for ensuring a stable—or even increased—supply of crude. Similar to a species' awareness of its most basic and essential relationship with a food source, political leaders by 1920 included petroleum supplies on the shortest list of critical priorities, essential to the security and future of the United States and other developed nations. Negotiations between nations now had to factor in this key logic and rationale, giving rise to one of the basic components of the concept of geopolitics.

GLOBAL EMERGENCE OF GEOPOLITICS DURING THE INTERWAR YEARS

In short, geopolitics emphasized nations' need for energy, particularly petroleum. Further agreements to manage oil markets were less inclusive, particularly excluding the United States, which feared they would violate the nation's antitrust laws. While the United States spent the early 1930s settling the public–private partnership that would administer its domestic supplies, the other Red Line nations secured their global spheres of petroleum influence. In 1934, Gulf and Anglo-Persian joined forces to develop Kuwait's oil. Not to miss out on the Middle Eastern prize, throughout the mid-1930s representatives of U.S. companies conducted testing in Saudi Arabia. In 1930, Standard Oil of California (SOCAL) had obtained a concession on the island of Bahrain, which lay off the coast of Saudi Arabia, and followed it with a concession in Saudi Arabia in 1933. Joining forces with a Texas company to form Caltex in 1936, SOCAL's claims in the area became the rationale for U.S. development of Saudi Arabia's oil. By the end of the decade, massive strikes had come in for American companies in both places.

Efforts by the United States and Britain to gain access and control to petroleum and other mineral resources was noted by other nations. Germany, Italy, and Japan, writes historian Eckes, "complained vigorously and repeatedly that the uneven distribution of these key materials was both inequitable and intolerable." In 1939, he continues, Britain and the United States controlled over three-fourths of all mineral resources and Germany, Italy, and Japan only 11 percent.[27] Although there are additional complications, Eckes argues that these deficiencies helped to drive these nations to consider warfare as a viable option to close the mineral gap. And, petroleum may have been the most significant of these minerals.

In the long-known oil regions of Baku, which was discussed above, Soviet occupation restored production in the 1920s and 1930s so that oil extraction in Azerbaijan in 1940 reached 22.2 million tons, which comprised 71.5 percent of the entire oil extraction in the USSR. During this period, the industry opened new fields, drained the Bibi Heybat Bay (1927), constructed the Baku-Batumi oil pipeline (1925), and drilled the first oil well in the open sea. These were monumental—even heroic—accomplishments for the pursuit of crude and demonstrated the growing geopolitical significance of crude between the World Wars. The strategic importance of energy resources helped to establish the points of conflict that were likely to cause future conflict and also foreshadowed the fashion in which wars would play out.

CONDUIT: Blitzkrieg: The Strategy of High-Energy Warfare

World War II left a number of indelible images on the human psyche, including concentration camps and the Holocaust as well as mushroom clouds and atomic technology. Another image, more prevalent among Europeans, was what the Nazi war machine did to modern warfare. Or, more precisely, what the Nazis integration of petroleum did to modern warfare.

Known as blitzkrieg or "lightning war," the Nazi technical innovations combined with a strategy to match. Gefreiter Mollmann, a German soldier, recalls the scene:

> The tanks roar ahead of us. It is a massive show of military might. They have smashed through all resistance, so that until now we have driven unopposed through enemy towns and villages. . . .
> The drivers are the silent heroes of this march. They clench their teeth. Stay awake at all costs! Roads, roads, roads—always the same. . . . We roll onwards, devouring the kilometres. . . . We advance—on and on. Past endless ditches by the roadside. For us there is a kind of "mystery" in roadside ditches. They characterize the nature of the advance! Abandoned vehicles, broken weapons, discarded ammunition of all kinds, helmets, uniform items. And everywhere the bodies of fallen enemies. We can see from the ditches exactly what happened here.[28]

Another soldier, Oberleutnant Dietz, adds: "There is no pause for rest: forwards—surprise attack—breakthrough—pursue, destroy and throw everything into it without keeping back even the very last reserves! Those are the mottoes in this war of fast-moving troops. The men's achievements are superhuman."[29] In fact, these accomplishments were only possible for humans who had mastered the use of petroleum and integrated it into their approach to battle.

The aggressive application on the battlefield of new devices, many of which were powered by burning gasoline, was an uneven revolution that had arrived rapidly in World War I and by World War II had crystallized into a new strategic approach to battle. The military journalist Lieutenant Werner Shafer, who moved across Europe with the invading Nazi forces, drew the distinction dramatically:

> The great hope of both officers and men is—the tanks! . . .
> Marching? We'll be racing. We're going to be asking a lot of our vehicles. . . . This Panzer army is fantastic! Like a creeping barrage it pushes forward at incredible speed toward the Marne. . . .
> We really only do measure distance in kilometres now. Our dynamic Panzer arm has spoilt us for any other kind of calculations. Would front-line soldiers in the Great War have believed that the [Nazi forces] could keep up this pace not only in the Polish blitzkrieg, but in France?[30]

Panzer tanks joined with motorcycle corps, autos, and trucks, as well as airplanes and rockets to use petroleum-based fuels to alter human warfare forever. As Nazi forces brought each of these innovations to the battlefield, the Allies sought to measure up and, ultimately, better them. And in the Pacific theater, Japanese warriors directed their imperialistic wishes through airplanes known as "zeroes" that could be delivered by massive aircraft carriers.

WORLD WAR II AS A WAR FOR OIL

By the late 1930s, crude had become a matter of urgency for any industrialized nation. While the search for oil was not yet a product of scarcity, existing reserves brought immediate interest from other nations. In the waning days of colonialism, oil reserves in the Middle East attracted the competing interest of many nations, none more than the immense—largely unclaimed—reserves of Saudi Arabia. As this struggle for influence unfolded, the world slid into World War II, what some scholars call a "war for oil." In particular, although the former Allied powers had locked up the known supply, geologists promised that Saudi Arabia's supply was beyond any comprehension. Diplomatic historian David Painter writes that American interest in Saudi Arabia was threefold: "growing concern over the adequacy of U.S. oil reserves, growing awareness of the extent of Saudi Arabia's oil potential, and growing fear that the British might use their influence in Saudi Arabia to the detriment of U.S. oil interests."[31]

In general, the need to secure crude became one of the great rationales for war among nations outside the power-brokering over Middle Eastern oil. Although the U.S. production of oil in 1941 stood at 64 percent of the world's supply, its consumption of energy had climbed at a similar rate. From 1920 to 1941, even though the American population grew from only 106 million to 133 million, annual consumption of oil and oil products increased from 4.3 barrels to 11.2 per capita. And, on the eve of the war, writes Painter, "the first 'Anglo-American petroleum order' had broken down under the impact of depression, world war, and the growing ability and desire of producing countries to control their economic destiny."[32] As each of the soon-to-be Axis powers confronted their own energy needs, they sought drilling agreements with Saudi Arabia by lavishing the king with gifts and offers of huge payments while also carrying out military expansion directed at controlling oil-producing nations.

Fuel for Japanese Imperialism

Particularly in the case of Japanese imperialism, the quest for oil fueled the drive toward World War II. The United States, which supplied approximately 80 percent of Japan's oil, joined other Western nations in condemning Japan's expansionist attacks on China and other neighbors. In 1937, President Franklin Roosevelt began publicly discussing an economic war, known as a "quarantine." With this threat being discussed openly, Japan tried to find ways of internally rationing its use of petroleum so that it would still be available if the United States refused it. On July 2, 1940, Roosevelt signed the National Defense Act after the Nazis invaded Western Europe. This policy provided him with the executive privilege to freeze economic activity to hostile nations. In retaliation, Japanese forces pushed deeper into Southeast Asia in order to access additional supplies of crude in the Dutch East Indies. As American leaders argued over whether or not to completely cut off Japan's supply in 1940–1941, Japanese leaders assumed it would occur and, therefore, feeling encircled by the Allied powers, devised a plan for a decisive attack that would bring the United States into the war. This attack was the naval and air assault on Pearl Harbor, headquarters of the navy's Pacific fleet.

Launched from aircraft carriers toward a distant, unsuspecting site, the Japanese attack on Pearl Harbor was made possible by the use of petroleum in warfare. Ironically, though, the devastation was not as significant to the U.S. Navy as it might have been with more careful, and thoughtful, planning. The Japanese attack destroyed a number of naval ships and nearly two hundred aircraft while killing more than twenty-four hundred U.S. service personnel; however, the U.S. fleet's stored supply of petroleum on Oahu remained largely unscathed. Some scholars have credited this to oversight; others suggest the opposite—that the Japanese hoped to put the supply to use in their own fleet following a full-scale invasion and occupation.[33] With a military strategy that depended on a ready supply of airplane fuel, such logic seems very sound.

Hitler's Blitzkrieg Guzzles Fuel

In Europe, few thinkers had identified the critical importance of fueling war machines more quickly than had Adolf Hitler, Germany's head of state. One historian refers to World War I as a static war and World War II as a war of motion.[34] Forming his government in 1933, the same year that Roosevelt came to power in the United States, Hitler had plans and ideas that knew no limits. A portion of his vision of an all-powerful Germany derived from what today we might call "energy independence." Synthetic fuels offered German scientists the ability to deliver new forms of energy to Hitler. These experiments, though, had begun earlier; in fact, some of them were carried out in the United States. IG Farben, the German chemical manufacturer, worked with Standard Oil to experiment with hydrogenation processes that synthetically derived additional fuel from petroleum. In 1931, the leader of Farben, Carl Bosch, shared the Nobel Prize in chemistry for the development of this important process.

With Hitler in power, such technical innovations became part of the German machine. Whereas the security importance of energy resources had led to close relationships between private oil companies and the governments in the United States and Great Britain, Hitler's Germany simply made IG Farben part of the Nazi state. In each situation, though, energy from petroleum was clearly identified as a matter of national security. Hitler expanded the application of synthetic fuels while also making the seizure of petroleum supplies a primary component of the imperialist expansion occurring after 1939.

His basic strategy of blitzkrieg ("lightning war") was organized around short, concentrated battles carried out over a broad area by mechanical forces. A portion of his desire for rapidity derived from a need to complete fighting before petroleum supplies ran out. Although petroleum supplies affected all of the Nazi commanders, it was the push of General Erwin Rommel across the wide expanse of northern Africa that most acutely demonstrates the importance of supply. His advance was forced to stop when his supplies ran out far from any German base. In hindsight, Rommel recalled: "The bravest men can do nothing without guns, the guns nothing without plenty of ammunition, and neither guns nor ammunition are of much use in mobile warfare unless there are vehicles with sufficient petrol to haul them around."[35] In addition, Hitler's revolutionary use of rocket technology for attacks on Europe used devices powered by synthetic fuels (primarily manufactured by slave labor). The realization of energy's importance had led him to create the most formidable military the world had ever seen—at least temporarily.

Fueling the Allied War Machine

In order to fight this war machine effectively, the Allies emphasized their greatest resources, including the courage of soldiers, the strategy of leaders, and, finally, a nearly endless supply of petroleum. This tremendous advantage was immediately transferred into strategy on the battlefield. At the peak levels of activity, American forces used one hundred times more gasoline in World War II than they had in World War I. Yergin estimates that the typical American division in World War I used 4,000 horsepower while in World War II this rose to 187,000 horsepower. In planning efforts of 1942, the U.S. Army estimated that each soldier in the field required approximately sixty-seven pounds of supplies and equipment to support him, of which half were petroleum products.[36] Between 1939 and 1943, U.S. oil consumption shot up by 28 percent.[37]

Although most of the Allies used American crude, the Soviets drew from the supplies in Baku that they had developed between the wars. Taking into consideration the growing requirements of petroleum in any war effort, the oil workers of Baku reached a record level of oil extraction in 1941—23,482 million tons. When thousands of oil workers had to leave the fields to fight in World War II, many of their industrial tasks were carried out by women. It is estimated that by the summer of 1942, twenty-five thousand women worked in the oil industry in Baku, accounting for approximately 33 percent of all Soviet workers. As the hub of the Soviet oil industry, Baku became a crucial target for Hitler.[38] On September 25, 1942, Nazi forces sought to gain control of this great supply of crude.

Prepared for evacuation, workers in Baku stopped production of 764 wells in 1942 and prepared to destroy them. Additionally, the industry sought to circumvent German blockades on the traditional modes of transportation by shipping the oil through Central Asia. For this purpose, the industry employed the first railway cisterns that were tugged afloat in the sea from Baku to Krasnovodsk. The Soviet State Defense Committee ordered the evacuation of approximately eleven thousand oil specialists and a great amount of equipment from Baku to Tatarstan, Bashkiria, and other regions of Russia in October 1942. Most of these workers were transported to the vicinities of the city of Kuybyshev (Tatarstan), where workers established "a second Baku. "Restoration of the industry began at the end of 1943 and after the Battle of Stalingrad, when the danger around Baku had passed. However, the war's impact was significantly reflected in Azerbaijan's level of oil extraction, which fell to 11.5 million tons in 1945."[39]

On the battlefield, petroleum made a dramatic difference during World War II. Actually, a significant change came *above* the battlefield. Using 100-octane gasoline, British Spitfires demonstrated a definite advantage over German Messerschmitt 109s (burning 87 octane) in air combat as early as 1940. Created in the United States and treated much like shipments of valuable gold, 100-octane gasoline was shipped through U-boat-infested waters under the strictest secrecy. Thus one of the greatest Allied resources for war was the advanced U.S. refining industry that utilized thermal cracking techniques to create fuels, such as 100-octane gasoline, that increased engine performance on the battlefield. No moment demonstrates this new strategic importance, though, as clearly as does the Allied invasion of Normandy. The delivery of fuel supplies presented a major component of the invasion. In fact, once the Allies moved through France, the difficulty facing General George Patton and others became that the fuel supply remained at the point of invasion, back in Normandy. The fast-moving Allied armies simply outran their supply line of petroleum. Yergin reports:

> Down to a half-day supply of gasoline, Patton was furious. He appeared "bellowing like an angry bull." At the headquarters of General Omar Bradley, commander of the American forces. "We'll win your goddam war if you'll keep Third Army going," he roared at Bradley. "Dammit, Brad, just give me 400,000 gallons of gasoline, and I'll put you inside Germany in two days."[40]

In the United States, which remained the world's largest oil supplier, Roosevelt forced the industry into new methods of organization, including systematizing distribution, standardizing products, and even the mundane such as using uniformly sized five-gallon gasoline cans—complete with revolutionary, built-in nozzles. To oversee these new standards and systems, President Roosevelt created the post of Petroleum Coordinator for Defense and moved Secretary of the Interior Harold Ickes to the task. One of Ickes's primary tasks was to alter the culture of those leading the petroleum industry, which was entirely organized under the concept of production surpluses. Appointed by letter, Ickes was assigned to:

> obtain . . . information as to (a) the military and civilian needs for petroleum and petroleum products and (b) any action proposed which will affect such availability of petroleum and petroleum products.[41]

Although his post focused on gathering information, the needs of the war effort forced him to openly discuss a new concept: conservation. Spring 1942, for instance, brought the first episodes of consumer rationing in the United States.

The rationing of World War II represents a direct link to the importance of petroleum for security purposes. Throughout the war, in which the United States was involved from late 1941 through August 1945, the scale and procedure for domestic gasoline rationing evolved. Initially, the use of gasoline for auto racing was banned throughout the United States. The next step came in May 1942 when rationing cards, which would be punched when a driver filled up at a gas station, were put into service in the eastern United States. Eventually, the cards gave way to the widespread use of coupons for gasoline purchasing; in general, though, rationing of gasoline focused on the more densely populated Northeast and allowed westerners and others to pump gasoline largely at will. Other measures implemented a thirty-five mile-per-hour speed limit and also sought to restrict "nonessential driving." This latter effort resulted in a rationing system of five separate grades of driving (essential, including doctors and clergy, to nonessential) to be demarcated by stickers that drivers displayed in their automobile windows. Typically, citizens at the basic level of nonessential driving received one to four gallons per week; essential drivers received no limit whatsoever. In addition, driving was influenced by other forms of rationing, including the conservation of rubber.

It is only a slight exaggeration to say that petroleum—and, by an overwhelming margin, that of the United States—enabled the Allied war machine to defeat the Axis powers. Although American consumers might have been largely shielded from this reality, U.S. leaders were not. In the halls of Washington, D.C., there was little doubt that any diplomatic strategy was required to at least consider the flow of and access to petroleum supplies.

NATURAL GAS AND PIPELINES ESTABLISH AN ENDURING ENERGY INFRASTRUCTURE

War also spurred the construction of critical energy infrastructure in the United States, particularly pipelines, which eventually enabled the commodification of

Moving energy in pipelines, whether oil or natural gas, emerged as crucial during World War II and remained so into the twenty-first century. LIBRARY OF CONGRESS PRINTS AND PHOTOGRAPHS DIVISION

the natural gas that had been evident but unrecoverable since the early oil wells. In the early 1900s, pipelines were leaky and they served best for local and limited regional delivery and were constructed in sections with overlapping, riveted ends and caulked. Electric arc welding improved pipeline technology during the first decades of the 1900s.

In the late 1800s, Polish engineers had developed an arc-welding system that used a carbon rod clamped in a wired handpiece to deliver current to the pointed end of the rod. A clamp held together the metal pieces to be welded and the operator scratched the point of the carbon rod against the metal to create a hot, conductive plasma that completed the circuit. By moving the rod along the line, the operator could melt it and fuse the metal together with a seal. While this method of welding worked for specific connections, such as to a boiler, it wasn't used in lieu of rivets on pipelines or ships until World War I. Historian Richard Rhodes writes that this change in ships occurred in the United States "to repair the many ships damaged or sabotoged by the Germans." In February 1918, U.S. Navy Secretary Josephus Daniels declared that the "British had faced so great a demand for oxygen and carbide for oxyacetylene welding that they had taken up electric welding as a substitute—and found it superior."[42]

Starting in 1925, the Magnolia Petroleum Company of Galveston, Texas, used arc welding to rebuild leaky pipelines that were previously built with rivets. Using this same technology, by 1931 pipeline workers were laying the first thousand-mile natural-gas pipeline from the Texas Panhandle to Chicago. For all of its advantages of cost and cleaner burning, natural gas remained limited by the cost of pipeline construction. These rates steadily dropped and by 1940, a national network of gas pipelines reached from Texas and Louisiana to the Midwest and into Pennsylvania.[43] Still, the high cost of shipping helped make gas expendable and huge volumes went to waste, either vented off or dumped. In fact, a report to Congress estimated that between 1919 and 1930, 20 percent more gas was wasted than consumed in the United States.

World War II spurred massive construction of oil pipelines throughout the United States in the 1940s, including the famous Big Inch and Little Big Inch, as an emergency undertaking of the federal government. At the war's close, these pipelines were closed and put on standby as interested parties developed their future usage. A vocal industry group argued for the pipelines to be reopened and dedicated to carry the waste natural gas from Texas to the eastern United States. With the energy transition playing out in real time, coal interests (including miners' unions) fought to keep out the gas that they thought would replace coal. In 1947, the Texas Eastern Transmission Corporation bought the Big Inch and Little Big Inch for transmission of natural gas, and they continue to function for this purpose today.[44]

CONCLUSION: EXTENDING THE GEOPOLITICS OF WORLD WAR II TO THE COLD WAR AND COLONIZATION

As World War II revealed petroleum's new importance, some of the most significant geopolitical shifts concerned regions with the largest reserves as nations competed to ensure their future supplies of energy. In particular, the mid-twentieth century witnessed a wholesale recalculation of the geopolitical importance of the Middle East. In the United States, for instance, a version of the wartime management of petroleum extended through the 1950s as Eisenhower's strategic materials commission. Thus, while the American public spent the post-war years basking in supplies of cheap petroleum—decadently finding ways of using it for even the most frivolous everyday tasks—military and political leaders began the process of fully inculcating the resource into the strategic planning of the nation's future. Specifically, they followed the model of Britain and France to increase their strategic connectivity in the oil-rich northern areas of the African continent. As early as 1943, Ickes had written: "We're Running Out of Oil!" He went on to stress that World War III would have to be fought with someone else's petroleum, because that of the United States was dwindling.[45]

The effort to reach out to King Ibn Saud and other regional leaders was carried out by government officials and American oil companies. Even during World War II, this effort was viewed by some American officials as a competition for favor with allies, particularly Britain. During the war, Ickes's efforts resulted in Roosevelt's creation of the Petroleum Reserves Corporation (PRC). In this experience, Ickes quickly learned the limits to involving the U.S. government directly in matters of the oil industry. In retreat, the emphasis of any federal initiative became forging diplomatic ties largely out of the view of the general public. These efforts, though, formed the relationships that would define U.S. relations and energy prominence into the twenty-first century—and they were not only with the governments of the Middle East.

Before the end of World War II and prior to Roosevelt's death in 1945, Churchill (now prime minister) chose to stare directly into the ongoing efforts by Ickes and others, it seemed, to compete with the British inroads to secure Middle East oil. On February 20, 1944, Churchill wired Roosevelt to say that he had been watching the telegrams from the United States about oil with "increasing misgivings." "A wrangle about oil," he continued,

> would be a poor prelude for the tremendous joint enterprise and sacrifice to which we have bound ourselves. . . . There is apprehension in some quarters here that the United States has a desire to deprive us of our oil assets in the Middle East on which, among other things, the whole supply of our Navy depends.[46]

He concluded that some British officials felt they were being "hustled" by the United States. Roosevelt did not back down, and their exchange went back and forth before finally concluding with mutual assurances that each nation would give way to the efforts of the other: British entered into Iran and Iraq and the United States into Saudi Arabia. The result was the Anglo-American Petroleum Agreement, which was signed on August 8, 1944, and assured the equity of all parties and the cooperative application of technology and developmental systems to extract petroleum from the Middle East and to bring it to the Allied powers. Facing unlikely approval in the Senate, the bill was pulled by President Roosevelt and made part of the negotiations in Yalta in January 1945.

Is it oversimplification to say that the world's energy future relied on the immense personal charm of its then dying president? Indeed, these deals were not just about FDR; however, despite the absence of a record of what transpired onboard the USS *Quincy* in February 1945, just after the Yalta Conference, there is no doubt that FDR set the course for a new relationship between these nations. As the ship sat in the Suez Canal Zone off Egypt, one ship brought Roosevelt and another brought Saudi king Ibn Saud. The existing record shows that the men bonded: the chain-smoking Roosevelt abstained in the king's presence; the king, injured at war and left with largely immobile legs, coveted Roosevelt's wheelchair and called the two men "twins." The two leaders talked for five hours about a Jewish homeland, the postwar configuration of the Middle East, and Saudi oil supplies. When Ibn Saud left the ship, Roosevelt sent with him his back-up wheelchair. Over the next half century, as American petroleum supplies diminished, Saudi Arabia became the nation's most trusted supplier.

Within twenty-four hours of Japanese surrender in August 1945, domestic gas rationing was ended in the United States. Consumers rebounded by designing their postwar lives around energy decadence, which will be discussed in the next chapter. During 1946, behind-the-scenes talks with Standard Oil of New Jersey and other companies worked to divvy up the opportunity to develop Arabian oil. By applying the concept of "supervening illegality" that Britain had used to seize shares and properties of any owner associated with the Axis powers, the oil executives voided the original agreement that had created the "red line" and divided up Middle Eastern oil prior to the war. Forcing their way to the table, American oil executives demanded to be officially cut into the Middle Eastern reserves. Although Aramco and British leaders eventually agreed to proceed with a new arrangement, France refused. When the parties approached Ibn Saud, possibly recalling his meeting with FDR at the close of World War II, he demanded only that the Americans be included. Ultimately, the group agreement of November 1948 laid out a new structure for the world—a geopolitical organization.

This new petroleum-centric worldview had taken shape during World War II. By 1941, Max Weston Thornburg, one of the vice presidents of the Bahrain Petroleum Company, had been brought into the U.S. State Department as an adviser. The seriousness of petroleum access is demonstrated by its growing importance to the State Department, and Thornburg worked to exactly this end through the war years. In addition to interacting with foreign competitors for supplies, petroleum diplomacy also was needed to manage a growing desire for resource nationalization in Mexico and Venezuela. Across the board, Thornburg argued that if the United States were to maintain its dominant position in world oil, "it would need a 'positive' foreign oil policy that protected its interests and anticipated problems between U.S. companies and foreign governments before they developed into crises."[47] In cases such as Venezuela, policies might require the United States to support political leaders who were more likely to work closely with American oil interests. Through the mid-twentieth century, such efforts, on the whole, proved more successful in Venezuela than in Mexico. Regardless, though, the U.S. place in a post–World War II world was obviously predicated on accessing critical energy resources.

The primary focus of this new world order remained the Middle East. U.S. State Department economic adviser Herbert Feis, who worked with Thornburg, noted of this moment in history: "In all surveys of the situation, the pencil came to an awed pause at one point and place—the Middle East."[48] Similar to a child's game of musical chairs, as the music stopped and each Western power paired up with oil-possessing regions or nations, the late-starting United States sat where no other nation was interested: Saudi Arabia.[49] Throughout 1943, amidst fear of British encroachment, the U.S. State Department used finances and diplomatic favor to lay the groundwork for its relationship with the Saudis. With the creation of the Petroleum Reserves Corporation in 1943, the United States made its task: "to buy or otherwise acquire reserves of proved petroleum from sources outside the U.S." This agency became the major mechanism for joining public–private effort that was needed to secure American energy interests in Saudi Arabia.[50]

In a lopsided and largely exploitative arrangement, petroleum companies associated with Allied powers used this moment in history to secure the world's precious supply of petroleum, and this agenda became a primary component of what became known as the "Cold War." In 1950 President Truman wrote to Ibn Saud: "I wish to renew to Your Majesty the assurances which have been made to you several times in the past, that the United States is interested in the preservation of the independence and territorial integrity of Saudi Arabia. No threat to your Kingdom could occur which would not be a matter of immediate concern to the United States."[51] Similar arrangements would eventually bring Kuwait into the American sphere of influence and also help to involve the United States in the internal politics of nations such as Iran.

NOTES

1. Peter Shulman, *Coal and Empire* (Baltimore: Johns Hopkins University Press 2015), p. 71.

2. Christopher Jones, Routes of Power (Cambridge: Harvard University Press, 2016), pp. 144–145, 196–197.

3. See Brian C. Black, *Crude Reality* (New York: Rowman & Littlefield, 2020).

4. Black, *Crude Reality*, pp. 38–39.

5. Shulman, p. 147.

6. Winston Churchill, *The World Crisis* (New York: Charles Scribner's Sons, 1923), p. 134.

7. Winston Churchill, House of Commons, June 17, 1914.

8. Leonardo Maurgeri, *The Age of Oil* (New York: Praeger, 2006), p. 24.

9. Maurgeri, pp. 41–42.

10. Maurgeri, p. 43.

11. Maurgeri, p. 45.

12. Rudi Volti, *Cars and Culture* (Baltimore: Johns Hopkins University Press, 2004), p. 44.

13. David Kirsch, *The Electric Vehicle and the Burden of History* (New Brunswick, NJ: Rutgers University Press 2000), pp. 162–164.

14. Volti, p. 46.

15. Kirsch, p. 165.

16. Daniel Yergin, *The Prize* (New York: Free Press, 2008), p. 168.

17. Yergin, pp. 170–172.

18. Yergin, p. 172.

19. Yergin, p. 177.

20. Yergin.

21. Yergin.

22. Painter writes: "[The] Red Line Agreement was to ensure that the development of the region's oil took place in a cooperative, rather than a competitive manner." The organization changed its name in 1929 to the Iraq Petroleum Company. Each organization was a British corporation and its legal provisions were enforceable in British courts. David Painter, *Oil and the American Century* (Baltimore: Johns Hopkins University Press, 1986), p. 183.

23. Brian Black, "The Most Important Road Trip in American History," *New York Times*, July 7, 2019. https://www.nytimes.com/2019/07/07/opinion/the-most-important-road-trip-in-american-history.html?smid=nytcore-ios-share.

24. Black, "Most Important Road Trip."

25. Black, *Crude Reality*, p.134.

26. Painter.

27. Alfred E. Eckes, *The United States and the Global Struggle for Minerals* (Austin: University of Texas Press, 1979), p. 58.

28. Heinz Guderian, *Blitzkrieg: In Their Own Words* (St Paul, MN: Zenith Books, 2005), pp. 76–77.

29. Guderian, p. 138.

30. Guderian, pp. 171–173.

31. Painter.

32. Painter, p. 9.

33. Yergin, p. 380.

34. Yergin, p. 382.

35. Yergin, p. 343.

36. Yergin, p. 382.

37. Painter, p. 34.

38. Yergin, p. 387.
39. Yergin.
40. Yergin.
41. Painter, p. 11.
42. Richard Rhodes, *Energy* (New York: Simon & Schuster 2019), pp. 260–261.
43. Rhodes, p. 263.
44. Rhodes, pp. 270–271.
45. Yergin, p. 395.
46. Yergin, p. 401.
47. Painter, p. 17.
48. Painter, p. 35.
49. Yergin, p. 423.
50. Painter, pp. 46–47.
51. Yergin, pp. 427–428.

6

Energy Technology and Empire
in the Cold War Era

GROUND: RED FOREST, UKRAINE, 2019

*Unmanned drones brought the unnerving reality back from the Ukraine to labora-
tories at the University of Bristol, UK: the ten-kilometer-square (four square miles)
Red Forest—a dense woodland of dead pine trees near the ruins of the old Chernobyl
nuclear reactor that weathered the brunt of the station's cloud of debris over thirty
years prior—remains among the most intense patches of radioactivity on Earth's
surface. The data left no doubt that the area will not be reinhabited for decades;
Ukraine authorities estimated that it may be thousands of years before the area could
be declared safe for human habitation.*

*The Red Forest and its surrounding region were transformed by an industrial
accident begun in the early hours of Friday April 25, 1986, during a test on the
Chernobyl 4 reactor prior to a routine shutdown. Unknown to the operators, the
reactor core was in an extremely unstable condition when they went to insert the control rods to shut down the reactor. As a result, there was a dramatic power surge that
caused explosions of steam that ultimately exposed the reactor core to the atmosphere.[1]*

*During the ensuing years, the Red Forest presented scientists with a full-size labo-
ratory in which to assess the impacts of radiation. Although no humans are allowed
to live in the extensive exclusion zones around the epicenter, animals and plants
still show signs of radiation poisoning. Scientists report that birds in these zones
have significantly smaller brains than normal and trees are growing more slowly. In
addition, they count many fewer insects, and game animals have shown dangerous
levels of radiation. Even more troubling, decomposers—organisms such as microbes,
fungi, and some types of insects that drive the process of decay—are lacking. This
was first indicated by the pine trees themselves, which turned red and died after the
accident; however, since 1991, scientific observers write that the dead trees have not*

decomposed normally. In their testing, in areas with no radiation, 70–90 percent of the leaves were gone after a year; but in places where more radiation was present, the leaves retained around 60 percent of their original weight. In short, the forest's basic function had broken down.[2]

With new knowledge rapidly expanding, systems were an important part of the modern world that emerged in the twentieth century. More and more world leaders understood that their nations' economic advancement required and depended on energy and the systems that were required to provide it. Supplies of resources that could provide such power for industry and everyday living became a strong predictor of global political standing. With these factors in mind, leaders of all sorts made energy supplies part of their social planning and political calculations. This is even true of superpowers—maybe even more so.

In recent years, scholars have steered beyond the posturing and weaponry of the Cold War to expose its foundation as a contest of ideologies between capitalism and communism. When deconstructed further, though, these ideas, writes historian Robert Marks, are better classified as "Consumerism versus Productionism."[3] The United States pursued a capitalist free-market economy that left it open to vacillations of boom or bust and individual investment. By contrast, the Soviet Union had a planned economy in which the central state made the decisions and choices about the allocation of resources, which helped make the economy immune to fluctuations. The most rapid period of industrial growth for the Soviet Union occurred just prior to World War II; however, after the conflict the Soviets focused on integrating the raw resources and productive capacity of Eastern Europe. Within this paradigm, each nation pursued a distinct economic development strategy. The approaches differed on many levels, including individual rights, markets, property ownership, and so forth; however, they shared the common element that each economy required a driver. Access to energy became an economic, geopolitical, and technological front for the conflict.

Particularly on the systems-level thinking of the modern human, energy requirements necessitated methods for assuring supplies that were either from elsewhere or within the nation. Later chapters will discuss the methods for maintaining fossil fuel supplies employed by both the Soviet Union and the United States. For each nation, however, the utopian, ideal system would be one that would not deplete and would actually allow energy to appear through more of an internal manufacturing model of production. In this systems level of thinking, each Cold War nation aggressively pursued power generated from nuclear reaction (which was part of the attraction that also drew many nonsuperpowers toward nuclear power as well). In this manner, as well as its relevance to weapons systems, nuclear technology became the true governing force behind the conflict that we call the Cold War. As each nation surged productivity to achieve its status as one of the world's two superpowers, it embodied the globe's most ad-vanced examples of the reality of the modern era that was discussed in chapter 3: in the twentieth century, national power required steady, reliable energy supplies.

Starting in the 1940s, a clearer understanding of energy's importance to na-tional development combined with a number of other factors to drive a selection of developed nations to commit to the future of nuclear power. Nuclear technol-ogy's technical and industrial needs were tailor-made for the era of the modern nation-state that followed World War II, particularly due to its complexity and potential danger; however, before we delve into these details, it is essential for readers of *THHN* to first recognize that by pursuing nuclear, these nations were also making a foundational national commitment that energy was an essential part of their future. Once strategic thinking moved out from this basic reality, the incredible technical challenges, extremely high investments, and catastrophic potential outcomes of nuclear all became palatable.

Finally, this seminal era grew from another basic reality of the conflict that became known as the Cold War: unlike any other energy technology, nuclear power never escaped its origins as a weapon. For more than half a century, the dual life of nuclear technology has simultaneously buoyed and restricted its ap-plications in the domestic sphere, varying with the latest news or accidents.

THE ATOMIC AGE BEGINS AS A BOMB

By the late 1930s, World War II threatened the globe. Leaders of every nation searched for any edge that would defeat the enemy forces. Scientists in the United States and Germany experimented with nuclear reactions that could be contained in a powerful bomb and applied in war. In Germany, leaders felt that such a technology might be a decisive force in the war effort. In reaction, U.S. scientists enlisted U.S. physicist Albert Einstein to write a letter about their research to President Franklin D. Roosevelt. In his letter Einstein stressed the technology's potential—particularly if it were developed by the enemy. In October 1939 Roosevelt authorized government funding for atomic research.[4]

Science and the military were linked in a way never before seen. However, first American scientists needed to demonstrate the viability of an atomic reaction. Of course, today the concept of force generated by separating atomic particles is fairly well known; however, in 1940 such a concept smacked of science fiction, even as both sides of World War II sought an advantage in applying it. In 1940 the U.S. physicists Enrico Fermi and Leo Szilard received a government contract to construct a reactor at Columbia University. Other reactor experiments took place in a laboratory under the west stands of Stagg Field at the University of Chicago. In December 1942 Fermi achieved what the scientists considered the first self-sustained nuclear reaction. It was time to take the reaction out of doors, and this process would greatly increase the scope and scale of the experiment.

Under the leadership of General Leslie Groves in February 1943, the U.S. military acquired 202,343 hectares of land near Hanford, Washington. This land was one of three primary locations of Project Trinity, which was assigned portions of the duty to produce useful atomic technology. The coordinated activity of these three locations under the auspices of the U.S. military became a path-breaking illustration of the planning and strategy that would define many modern corporations. Hanford used water power to separate plutonium and produce the grade necessary for weapons use. Oak Ridge in Tennessee coordinated the production of uranium. These production facilities then fueled the heart of the undertaking, contained in Los Alamos, New Mexico, under the direction of the U.S. physicist J. Robert Oppenheimer.[5]

Oppenheimer supervised the team of nuclear theoreticians who would devise the formulas using atomic reactions within a weapon. Scientists from a variety of fields were involved in this complex theoretical mission. After theories were in place and materials delivered, the project became one of assembling and testing the technology in the form of a bomb. All of this needed to take place on the vast Los Alamos compound under complete secrecy. However, the urgency of war convinced many people that this well-orchestrated, corporate-like enterprise was the best way to save thousands of U.S. lives.[6]

By 1944 World War II had wrought terrible destruction on the world, and the European theater of war would soon close with Germany's surrender. Although Germany's pursuit of atomic weapons technology had fueled the competitive efforts of U.S. scientists, German surrender did not end the U.S. atomic project. The Pacific theater of war remained active, and Japan did not accept offers to surrender. Project Trinity moved forward, using the Japanese cities Hiroshima and Nagasaki as the test laboratories of initial atomic bomb explosions. The U.S. bomber Enola Gay released a uranium bomb on Hiroshima on August 6, 1945, and the U.S. bomber Bock's Car released a plutonium bomb on Nagasaki three days later. Death tolls vary between 150,000 and 300,000, and most were Japanese civilians. The atomic age, and life with the bomb, had begun.

Atomic Technology as a Tool to Define the Nation's Futures

Experiments and tests with nuclear and hydrogen bombs continued for nearly twenty years after World War II. Many of the scientists who worked on the original experiments, however, hoped that the technology could ultimately have nonmilitary applications. Oppenheimer eventually felt that the public had changed its attitude toward scientific exploration because of the bomb. "We have made a thing," he said in a 1946 speech, "a most terrible weapon, that has altered abruptly and profoundly the nature of the world . . . a thing that by all the standards of the world we grew up in is an evil thing."[7]

Many of the scientists involved believed that atomic technology required controls unlike those of any previous innovation. Shortly after the bombings a movement began to establish a global board of scientists who would administer the technology with no political affiliation. However, wresting control of this new tool for global influence from the U.S. military proved impossible. The U.S. Atomic Energy Commission (AEC), formed in 1946, placed the U.S. military and governmental authority in control of the weapons technology and other uses to which it might be put. With the "nuclear trump card," the United States catapulted to the top of global leadership. In the United States, the Atoms for Peace program emphasized that it was impossible for a nuclear plant to behave as a bomb would—that it could not explode. The Atomic Energy Commission, which succeeded the Manhattan Engineering District as of January 1, 1947, did much to encourage the commercial use of nuclear reactors for the generation of electricity. Lewis L. Strauss, chair of the AEC, proclaimed that nuclear power would soon be "too cheap to meter." Efforts began immediately to disassociate atomic technology from weapons and to ally it more with energy production.

Regardless of the output, nuclear reactions required the rapid acquisition of scarce high-grade uranium ore, which prospectors sought throughout the American West and elsewhere, particularly South Africa. With the Atomic Energy Act of 1946, the United States made development of the nuclear energy technology the monopoly of the federal government, and any development was considered a classified secret. One historian, writes Rhodes, referred to the AEC as "the most totalitarian governmental commission in the history of the country."[8] Its goal of "domesticating the atom" led the AEC and other organizations to sponsor a barrage of popular articles concerning a future in which roads were created through the use of atomic bombs and radiation was employed to cure cancer.[9]

In the media the atomic future included images of atomic-powered agriculture and automobiles. There were optimistic projections of vast amounts of energy being harnessed, without relying on limited natural resources like coal or oil. For the administration of U.S. president Dwight Eisenhower the technology meant expansion of U.S. economic and commercial capabilities with unlimited supplies of electricity. McNeill and Engelke write that the Cold War "justified . . . heroic commitment of money, labor, and planning to gigantic state-sponsored infrastructure projects and development campaigns."[10] Particularly in productionist-oriented economies such as the Soviet Union and China, the Cold War stimulated and rationalized schemes for enhancing "economic self-sufficiency," and unlimited power was its centerpiece. In his famous 1953 address to the United Nations, President Eisenhower specified that while nuclear technology would have many applications, its special purpose would be "to provide abundant electrical energy in the power-starved areas of the world."[11]

Chapter 6

CONDUIT: Fission Reactors and Shippingport, Pennsylvania

Creating electricity from the fission reaction is a fairly simple process. Similar to power generators fueled by fossil fuel, nuclear plants use the heat of thermal energy to turn turbines that generate electricity. The thermal energy comes from nuclear fission, which is made when a neutron emitted by a uranium nucleus strikes another uranium nucleus, which emits more neutrons and heat as it breaks apart. If the new neutrons strike other nuclei, chain reactions take place. These chain reactions are the source of nuclear energy, which then heats water to power the turbines.

Under the sponsorship of the AEC, initial reactors produced enough power to light only four 150-watt light bulbs; however, the technology was expanding rapidly. The Experimental Breeder Reactor (EBR I) was constructed at the National Reactor Testing Station near Idaho Falls, Idaho, by the Argonne National Laboratory of the University of Chicago, the lab that began with Enrico Fermi's pile on the university squash court. The next day, EBR I was powered up and produced 100 watts of power. The Argonne Lab also designed BORAX III, the first reactor to provide an entire town's electric power, which began producing power for Arco, Idaho, on July 17,

Shippingport, Pennsylvania, was the first American attempt to use nuclear power as a public utility. LIBRARY OF CONGRESS PRINTS AND PHOTOGRAPHS DIVISION

1955. Interestingly, connecting nuclear-generated electricity to an existing grid, however, began in a tiny test near Moscow in 1954 and quickly grew in larger examples in the United Kingdom and the United States in 1956–1957.[12] During these various experimental efforts, Admiral Hyman Rickover, who had first been assigned to the Oak Ridge, Tennessee, facility in 1946 and then focused his efforts on developing a navy program for nuclear submarine propulsion, determined to change the form of the fuel from uranium metal to uranium dioxide, a ceramic.

The pilot project to apply this technology for power generation was carried out by the Duquesne Light Company in Shippingport, Pennsylvania, in 1957. The 60-megawatt breeder reactor plant, which opened in 1957, was designed by Westinghouse but developed by Rickover. Ground breaking for the project was held on September 6, 1954, which Rhodes describes in this fashion: "Waving a magic wand—a neutron source—over a transmitter in Denver where he was recovering from a heart attack, Eisenhower activated a robot bulldozer in Shippingport to turn the first dirt for the new power plant."[13] In order to distinguish it from Soviet reactors, the AEC described Shippingport as "the world's first full-scale atomic electric plant devoted exclusively to peacetime uses."[14]

Shippingport was part of a federal power demonstration program designed to stimulate the construction of nuclear power plants by private utilities, and it remained operational until 1982. Most important, writes Rhodes, Rickover had early on "made the historic decision to moderate his submarine and large-ship reactors with water rather than a less familiar but more efficient coolant such as liquid sodium." The choice would "reverberate through the years," and other countries made their own choices, such as heavy water, helium, sodium, lead, or, at Chernobyl, graphite moderation with water cooling.[15]

Industry seemed reticent, however, and no plants were immediately ordered despite the AEC's encouragement and research. Government influence was deemed essential, and private companies were only interested if significant assistance came from the federal funds in the form of research and development assistance and waivers of fuel inventory charges for the first five years of plant operation. Government leaders in the United States intended that its assistance would only be temporary. And in 1957 Chairman Strauss stressed that if industry did not seize the opportunity to build plants within a reasonable time, the commission would take steps to build the reactors on "its own initiative."

Cold War and Nuclear Innovation

While atomic knowledge unmistakably altered military strategy, its interplay with Cold War economics particularly grew out of its possibility as an almost unlimited source of energy. Historian Marks writes: "More than anything else, the Cold War was a battle of economies."[16] He explains that the United States emphasized consumerism and the Soviet Union focused on productionism. As symbols of the modern era, each economic model, in fact, lent itself to the possibilities of developing nuclear power; however, each model can now be seen to have carried very different outcomes.

In the United States, government subsidy and regulation guided the development of nuclear power; however, it remained one source of electricity in a competitive marketplace. Initially, this connection to positivist "atomic culture" provided a source of American pride and made each community yearn to have plans for its own nuclear power plant after World War II. In general, this consumerist approach also left the nuclear power industry susceptible to changes in market preferences.

By contrast, the productionist approach allowed for a top-down method of implementation that has allowed the nuclear industry to first become a strong presence in the Soviet Union and then throughout the globe in political environments that grew either from authoritarianism or socialism. Particularly in the Soviet Union, this approach often led to more lax enforcement of regulation and safety measures. Similar to automobiles, writes McNeill,

> atomic power had its origins in European science, reached maturity in the U.S., and subsequently spread unevenly around the world. The world's first self-sustaining nuclear reaction took place in 1942 in a squash court at the Univ of Chicago with U.S. search to build a weapon. Civilian nuclear power started up in 1954 in the USSR, 1955 in the UK, and 1956 in the U.S. Nuclear power held some of the same political attraction as dam building: it signified vigor and modernity.[17]

As the Cold War took shape around nuclear weapons, the Eisenhower administration looked for ways to define a domestic role for nuclear power even as Soviet missiles threatened each American. In the United States, Project Plowshare grew out of the administration's effort to turn the destructive weapon into a domestic power producer. The list of possible applications was awesome: laser-cut highways passing through mountains; nuclear-powered greenhouses built by federal funds in the Midwest to enhance crop production; and irradiated soils to simplify weed and pest management.

When the first nuclear weapons exploded over Japan in 1945, observers all over the world knew that human life had changed in an instant. In the years since, nuclear technology has struggled to define itself as a public good when the public seemed more inclined to view it as an evil. Its proponents argue that electricity made from nuclear reactors has the capability to power the world more cleanly than can any other resource. Opponents are less sure. As the debate rages, nuclear power has become an increasingly important part of regional dynamics: it can liberate nations from the need to trade in fossil fuels and, thereby, acquire inexpensive power that might advance society in important ways. Correspondingly, other nations in Africa elect to receive nuclear waste from France and elsewhere, which could ultimately create health implications for an entire region. As an energy source or as a weapon, atomic technology has remained one of the most important transborder issues for almost a century.

Throughout the world, but particularly in the United States and Soviet Union, nuclear technology benefited from enormous government subsidies—including limits on regulations and insurance rates—that would help to stimulate rapid development. By 1998, twenty-nine countries operated some 437 nuclear power plants. Between 1965 and 1980 global electricity generated from nuclear power grew from less than 1 percent to 10 percent, peaking at 13 percent in 2013. As McNeill observes, though, "No nuclear power plant anywhere made commercial sense: they all survived on an 'insane' economics of massive subsidy."[18] On the world stage, nuclear power became the answer to desires for rapid national development for nations that lacked access to adequate supplies of fossil fuels. By 2010, such strategic thinking directly influenced a variety of nations to emphasize nuclear generated electricity: approximately half of electricity in France, Lithuania, and Belgium; a quarter in Japan and South Korea; and a fifth in the United States. Obviously, an acknowledgment of national needs for power influenced the future of nuclear energy in a transborder fashion.

Although domestic power production, with massive federal subsidies, would be the long-term product of government efforts, the atom could never fully escape neither its military capabilities nor its possible dangerous outcomes. This was most clear when nuclear power plants experienced industrial accidents.

Chernobyl Power Plant, Ukraine, April 25, 1986

Under normal operation, nuclear power plants marked a great advancement in the production of electricity—a seemingly ideal source of power; however, the technology always possessed significant potential difficulties and dangers. On April 25, 1986, Soviet reactor number 4 at the Chernobyl power plant in Ukraine (then part of the Soviet Union) realized these dark possibilities. Decades later, an entire region—as well as nuclear power in general—has not yet recovered, which was exemplified by the condition of the Red Forest, described above.

The problems at Chernobyl began with its design: it had no containment structure and was operated by poorly trained technicians. In a terrible combination of factors, worst-case scenarios were compounded. On that day, technicians shut down several emergency systems in order to test how the reactor might respond in trying conditions. Unbeknownst to them, the backup cooling system (which should have been continually operating in order to prevent a full core meltdown) was not functioning. As their test continued, heat surged uncontrollably. When it made contact with the water in the reactor, it detonated a massive explosion—forceful enough to lift off the reactor's 1,000-ton roof—that then released radioactive materials into the surrounding region with enough force that they also extended high into Earth's atmosphere.

A man-made, out-of-control volcano of toxic spew, Chernobyl fires were of immediate danger and were fought by thousands of fire and military professionals (most of whom suffered dire immediate or long-term circumstances) while also being dowsed by various materials dropped from helicopters. Of course, though, the radioactivity that was generated by the accident was so significant that the Soviets evacuated an entire city and region (more than 130,000 residents in total). Most of this area remains uninhabitable decades later. Incapable of being remediated, the site of the reactor ultimately had to be partitioned by a 35,000-ton, 20-meter-thick concrete and steel structure. Although challenging to document, the Chernobyl accident is referred to as "the greatest technological catastrophe in human history," killing several thousand Soviets immediately and very likely millions in the region when one considers the cancers caused by the radioactivity.[19]

CURRENT: Accidents Fuel Public Doubt

A number of nuclear power plant accidents occurred before the late 1970s, but they went largely unnoticed by the American public. Nuclear power became increasingly popular, even though critics continued to argue issues of safety. Nuclear power plants differ in numerous ways from other sites of power generation, but most important might be the concentration of activities—including waste products—that remain in a single site and, thereby, heighten the possibility of a breakdown in the system. Therefore, it was only a matter of time until nuclear technology, with its tenuous reactions and toxic by-products, faced a very serious public reckoning.

First, in 1979 the United States experienced its first nuclear accident in a residential area outside of Harrisburg, Pennsylvania. The accident at Three Mile Island (TMI) nuclear power plant entirely altered the landscape of American power generation. Although involving only a relatively minor release of radioactive gas, this accident demonstrated the public's lack of knowledge. Panic ripped through the state, and Harrisburg was partially evacuated.

While the international community took notice of the TMI accident, it clearly did not present a grave threat to the world. The world's other superpower had even greater difficulty with its atomic industry, which was plagued by accidents throughout this era. None, however, compared to the Chernobyl meltdown, which is believed to have released thirty to forty times the radioactivity of the atomic bombs dropped on Hiroshima and Nagasaki. By the early twenty-first century, the demand for electricity continued to fuel interest in developing nuclear power with China leading the way. For many nations, the great promise of nuclear outweighed its possible dangers. By 2010, 440 nuclear power plants were in operation spread among forty-four countries.

Into nuclear power's great future, however, came one other well-known event in 2011 at the Fukushima Daiichi reactor in Japan. On March 11, 2011, an earthquake led to a tsunami event that brought tides that overwhelmed coastal areas of Japan, including the nuclear reactors at this plant. Of the six reactors at the site, three are known to have experienced full meltdowns. The immediate emergency brought great confusion and, ultimately, wholesale evacuation of a twenty-kilometer area. Within months, Japan had shut down all fifty-four of its reactors. Although a few were returned to service, public confidence in nuclear has not returned, and the gap in electricity production has been made up by the use of fossil fuels.

Clearly, the Fukushima disaster focused attention on the need to properly position reactors; however, for many observers, the event presented additional evidence that this technology was not ready for domestic use. Could nations ever feel confident that they had considered every possible variable that could befall a nuclear reactor?

NUCLEAR POWER PLANTS DOUBT IN THE DEVELOPMENT IDEAL

The implications of nuclear weapons and nuclear power had already been of great interest to environmental organizations before Chernobyl. After Chernobyl, international environmental organizations such as Greenpeace dubbed nuclear power a transborder environmental disaster waiting to happen. Interestingly, even within the environmental movement, nuclear power maintained significant support due to its potential for generating "clean" power. Whereas almost every other method for producing large amounts of electricity creates smoke or other pollution, nuclear power creates only water vapor. Yet, at least in the public's mind, there remained the possibility of atomic explosions.

Although accidents decreased the American domestic interest in nuclear power generation, the international community refused to be so quick to discount the technology. Since the early 1990s nuclear power had become one of the fastest-growing sources of electricity in the world. In the first decade of the twenty-first century, nations that depended on nuclear power for at least one-quarter of their electricity included Belgium, Bulgaria, Hungary, Japan, Lithuania, Slovakia, South Korea, Sweden, Switzerland, Slovenia, and Ukraine. As concerns grew for the need to account for carbon emissions, nuclear remained appealing as an alternative to fossil fuels to many nations.

Clearly, though, regardless of the use to which it is put, nuclear energy continues to be plagued by its most nagging side effect: even if the reactor works perfectly for its service lifetime, the nuclear process generates dangerous waste. In fact, reactor waste from spent fuel rods is believed to remain toxic to humans for fifty thousand years. At present each nuclear nation makes its own arrangements for the waste. U.S. nuclear utilities now store radioactive waste at more than seventy locations while they await the fate of the effort to construct and open a nuclear waste repository inside Nevada's Yucca Mountain. This decades-long controversy was most recently defunded (in 2010) on the federal level, leaving the United States with no specific strategy for its waste storage. Internationally, the situation is not much clearer. Opponents in Germany have obstructed nuclear waste convoys, and shipments of plutonium-bearing waste to Japan for reprocessing are often placed under dispute. Some observers have voiced concern that less developed nations will offer themselves as waste dumps for the more developed nations. The income from such an arrangement may be too much to turn down for many nations.

Rising energy prices of all sorts, however, brought the nuclear industry robust interest in the twenty-first century. In particular, new attention has been focused on reprocessing used nuclear materials. In this fashion, the very idea of "nuclear waste" has been redefined. Proponents argue that there is no such thing as waste if the spent rods can be reprocessed in order to fuel other types of plants. These efforts are particularly advanced in France, the global leader in nuclear power. In the energy industry, many observers continue to believe that nuclear power remains the best hope to power the future. The issues of safety and waste removal need to be dealt with. However, in nations with scarce supplies of energy resources, nuclear power—even with its related concerns—remains the most affordable alternative and possibly the most sustainable energy for the future.

As McNeill sums up, atomic weapons programs of the Cold War era killed a few hundred thousand or at most a million humans; however, due to the dangers of waste generated from nuclear reactions in general, this is only a fraction of the story. The story doesn't stop with weapons, writes McNeill:

> It will not end for at least one hundred thousand more years. Most radioactivity decays within hours, days, or months and quickly ceases to carry dangers for living creatures. . . . [However, some wastes] remain lethally radioactive for more than a hundred thousand years. This is a waste management obligation bequeathed to the next three thousand human generations. If not consistently handled adroitly, this will elevate rates of leukemia and certain cancers in humans, especially children, for a long time to come. . . . [P]eople will either manage Cold War nuclear wastes through all the political turmoil, revolutions, wars, regime changes, state failures, pandemics, earthquakes, megafloods, sea level rises and falls, ice ages, and asteroid impacts that the future holds, or inadvertently suffer the consequences.[20]

As a representation of the Cold War ideologies of each superpower, nuclear power can also be credited with the end of the conflict—and of the Soviet Union—in the 1990s. As McNeill writes, Chernobyl and the effort to cover it up that followed "knocked one of the last props out from under the Soviet Union. It completely changed the public perception of nuclear power plants around the world, but especially in Europe, making it politically unpalatable except in a few countries."[21] In stark contrast to other energy sources, McNeill argues:

> Nuclear power did not replace other forms of energy production, as the car did the horse. It did not find companion innovations, technical and social, to form a new cluster that would remake the world, the way oil and ICE had done. Instead, nuclear power complemented fossil fuels; it never accounted for more than 5 % of world's energy supply. . . . No single technology, not even nuclear power, matched the Motown cluster in its capacity to alter both society and nature.[22]

MANAGING THE POWER OF RIVERS FOR THE COLD WAR CAUSE

Hydroelectric dams also became instrumental tools for economic development within each superpower. American efforts in the West and South were discussed earlier, particularly due to their military applications. In the USSR, the first major installation occurred along the Volga River in 1937 and others followed along the Dnieper, Don, and Dniester Rivers. By the 1950s, diversion of some sort reduced the streamflow of all large rivers in the southwestern USSR and one of the lasting impacts of Soviet development emerged as impacts on the Caspian and Aral Seas.

The Aral was affected by Soviet efforts after 1950 to manipulate Central Asia's greatest rivers, the Syr Dar'ya and the Amu Dar'ya. Though primarily for irrigation, these efforts, writes McNeill, constituted a "planned assassination." Massive investment in cotton growing organized the efforts that dried up the Aral, and by the mid-1990s only one-tenth of its former water influx arrived to the sea—sometimes none at all. Due to this alteration, the salinity level of the remaining sea tripled between 1960 and 1993. Once referred to as the "Blue Sea," experts expect the Aral Sea will be a salt pan of lifeless, brackish water.[23] Despite these outcomes, when the USSR collapsed in the 1990s it was forced to abandon even larger ambitions to reverse the flow of the great Siberian rivers, Ob and Yenisei.

The awareness of energy centrality also influenced the exportation of hydroelectric technology in hopes of gaining the allegiance of developing nations. Between 1930 and 1970, dam building was pursued by ambitious, modernizing states, particularly colonial and newly independent nations. The technology itself became a way for either the United States or the Soviet Union to display the virtues of their social and political systems. The dams' political utility, writes McNeill, "helps explain why so many uneconomic and ecologically dubious dams exist. On average, during the 1960s more than one large dam was completed per day.[24]

FOSSIL FUEL DEVELOPMENT IN THE COLD WAR

Despite these energy initiatives, the power structure of the Cold War remained firmly based in fossil fuels. While the United States focused on opportunities in the Middle East, Soviet officials expanded Baku's production. One of the greatest efforts focused on creating the world's first deep, off-shore wells near Neft Dashlari ("Oil Rocks"), which lay approximately 110 kilometers from Baku on the Caspian Sea. On November 14, 1948, the first troop of oil workers headed by Nikolai Baibakov landed in the open sea on a group of rocks called Gara Dashlar ("Black Rocks"). In his troop was the geologist Agagurban Aliyev, who was the author of the idea that there is oil in the sea.[25] After constructing a small electric power station on the rocks, they started drilling the first well on June 24, 1949, and it came in on November 7 at a depth of 1,100 meters. Within a few months, tankers had begun taking the oil to shore. Soon, an island was constructed on the site of Oil Rocks, and then another. In 1952, piers were added to connect each of the artificial islands. Soon, Oil Rocks received five- and even nine-story buildings, including hostels, hospitals, palaces of culture, and bakery factories; a lemonade workshop was constructed, and a park with trees was created. Since 1949 nearly two thousand wells have been drilled on Oil Rocks, and its production is more than 160 million tons of oil and 12 billion cubic meters of gas.

Additionally, the Soviets built the world's longest pipeline during the 1960s. The Druzhba, which is also referred as the Friendship Pipeline and the Comecon Pipeline, extends 4,000 kilometers (2,500 miles) from southeast Russia to points in Ukraine, Belarus, Poland, Hungary, Slovakia, Czech Republic, and Germany. Binding together the Soviet empire since 1964, the pipeline carried Russian crude to the energy-hungry western regions of the Soviet Union, as well as to "fraternal socialist allies" of the former Soviet bloc and to Western Europe. Today, it is the largest principal artery for the transportation of Russian (and Kazakh) oil across Europe.

Despite this increasing infrastructure, fluctuation in global oil prices and in Soviet oil production during the 1980s is believed by some scholars to have contributed to the collapse of the Soviet Union by the end of the decade. During the 1970s, the Soviet Union weathered the first oil shocks relatively unscathed by maintaining its own stable supply of crude. However, the Soviet price of oil had been artificially set much lower than the world price and much lower than its scarcity value within the communist system. This low price, coupled with virtually unlimited supplies up to the 1980s, subsidized the Soviet and Eastern European economies. However, Soviet oil production dropped by approximately 30 percent between 1988 and 1992 and the internal price needed to increase. This pressure created an oil crisis within the Soviet Union, and much of the export of oil to Eastern Europe ceased. Therefore, production decreased first and then consumption declined, which forced conservation and high prices—neither of which could be supported by Soviet and Eastern European economies. This internal economic decline, argue some scholars, played an important role in the Soviet Union's inability to compete globally, which destabilized the bipolarity that had defined the Cold War.[26]

Whatever the cause of Soviet decline, the end of the Cold War brought new opportunities for Baku oil, primarily in the form of new pipelines. The best known is the Baku-Tbilisi-Ceyhan pipeline that extends 1,768 kilometers (1,099 miles) from the Azeri-Chirag-Guneshli oil field near the Caspian Sea to the Mediterranean Sea. The pipeline provided Azerbaijan with great support of its new autonomy by tying Baku, the capital of Azerbaijan, with Tbilisi, the capital of Georgia, and Ceyhan, a port on the southeastern Mediterranean coast of Turkey. Today, it is the second longest oil pipeline in the former Soviet Union after the Druzhba. Although the pipeline is only partly in the former Soviet Union, it has delivered crude to Ceyhan since 2006.

This pipeline and the immense amount of oil that it could bring influenced global politics immediately. The United States and other Western nations have become much more involved in the affairs of the three nations through which oil flows. The countries have been trying to use the involvement as a counterbalance to Russian and Iranian economic and military dominance in the region. In recent years, Chinese influence has also grown. In short, after the Cold War, the Baku oil supplies have made these remote nations important power brokers on the world stage.

A WAR OF IDEOLOGY: MODELING CONSUMPTION

From the great kitchen debate between superpower leaders to the everyday life of American citizens, the emergent details of Americans' high-energy existence became symbols of the success of consumer society. Similar to the use of petroleum for transportation, the basic integration of petroleum into these other aspects of human life grew from a few basic priorities, including flexibility, planned obsolescence, and disposability. Cheap oil often helped humans to make cheap things, which appealed to mass consumers and helped to fuel wide-scale changes, ranging from product packaging to large, big-box stores like Walmart. At other times, cheap oil allowed chemists to derive cheap replication of costlier products, usually made from polymers—what we know of as plastics.

These plastics are suggestive of a larger pattern of commodities deriving from petroleum referred to as "synthetics," which might best be considered inexpensive replicas (whether on the level of chemical compounds or the products that these elements are used to create). Most of us know to describe plastic as a "petroleum by-product," but few of us know what role black gold plays in its production. In fact, for its early decades, plastics required no petroleum. The creation of synthetic materials that are related to plastics began in 1907 when a New York chemist named Leo Baekeland developed a liquid material that when cooled hardened into a replica of whatever form one chose. He called this resin material Bakelite. This new material was the first thermoset plastic, which meant that it would not lose the shape that it had taken. In fact, Bakelite would not burn, boil, melt, or dissolve in any commonly available acid or solvent.

In this same general product genre, inventors in the early 1900s developed products such as rayon and cellophane. Very often, large chemical companies such as DuPont had researchers constantly working in labs to develop any synthetic material that might prove to be useful. In this fashion, DuPont developed nylon during the 1930s and the first pair of stockings in 1939. Many similar innovations also occurred in the 1930s, including polyvinyl chloride (PVC), vinyl Saran Wrap, Teflon, and polyethylene. Although each of these items possessed well-known domestic uses, most of them were first used in other substances. For instance, during World War II polyethylene was used first as an underwater cable coating and then as a critical insulating material in radar units. By decreasing the weight of radar units, this material made the technology more portable so that it could be placed on planes.

During the 1930s, it became a sign of progress (breaking from the past) to insert these obviously man-made objects of modernity into the most mundane locations in our everyday life. Still, most of these objects were not ubiquitous. Their limited production most often began with coal, from which chemists rent phenol. The process of polymerization resulted in a resin that was formed by condensing phenol and formaldehyde. This resin, then, could be shaped and colored for whatever purpose was desired. With this process, chemists created a string of commercial products continuing from the early Bakelite, including celluloid, which ultimately led to acrylic plastics and an array of vinyl compounds and ultimately to polystyrene.

Historian Jeffrey Meikle notes that these developments meant that by the end of World War II, plastic had become cheaper, less durable, lighter, and increasingly more plentiful. The evolution led to a new category known as thermoplastics, which was "driven not so much by market demand as by the pressure of supply, an overabundance of chemical raw materials, waiting to be exploited."[27] The primary substitute, of course, was petroleum, which could create derivatives similar to those gotten from coal. Once again, the key to expansion was cheap oil.

By 1976, more plastic was manufactured, in terms of cubic volume, than all steel, copper, and aluminum combined. In part, the proliferation of plastics stemmed from the idiosyncrasies of production. Relative to other products, plastics are expensive to manufacture in small quantities because of the high fixed costs involved in making the molds and production equipment. Therefore, companies must produce huge quantities to recoup their investment—a situation tailor-made for conspicuous consumption. Today, five resins account for nearly 60 percent of all plastics used by consumers: low-density polyethylene, used in garbage bags; polyvinyl chloride, used in cooking oil bottles; high-density polyethylene, used in milk jugs; polypropylene, used in car battery cases; and polystyrene, used in disposable food containers. With so much plastic in our lives, we have learned that the material has another attribute beyond its flexibility of form: it is remarkably durable. In fact, it is almost impossible to dispose of!

Although residents of less developed countries have only slowly increased use for plastics, they can feel some of the impact of their use on the other side of the gap. Nondegradable plastic packaging is blamed for filling commercial landfills, increasing their operational expense, contaminating the environment, and posing a threat to animal and marine life. Together, this plastic waste accounts for about one-quarter of all municipal solid waste. This is a particularly big problem because of plastic's remarkable durability. By the 1980s, an estimated fifty-two million pounds of packaging were being dumped from commercial fleets into the ocean every year in addition to three hundred million pounds of plastic fishing nets. These trends helped to create one of the most bizarre global symbols of plastics and the era of conspicuous consumption: tidal accumulations of plastic trash in terribly remote locations.

During the twentieth century, the domestic environment of humans in developed countries became one of the greatest symbols of the gulf between societies. In these new forms, access to cheap petroleum generated important distinctions in comfort, safety, and health among humans. In terms of human homes, developed societies used petroleum-powered transportation to decentralize their living environment into variations of the suburban housing development. Heavy machinery similar to that which transformed agriculture also homogenized topography for use in creating housing tracts. In each of these homes, prefabricated materials, often enabled by the petrochemical industry, allowed the prices of secure housing to drop considerably. All over the world, new housing became accessible to new groups of humans. Within many of these homes, humans lived with comfort and safety previously unknown to most of the species—particularly in hostile climates.

Similar to petroleum's role in plastics, petrochemical "feedstocks" are used to produce plastics, drugs, detergents, and synthetic fibers. The chemical industry alone uses almost 1.5 million barrels of natural gas liquids and refinery gases a day as feedstocks. These feedstocks are obtained from processing various petroleum fuels and reducing them to their basic chemical elements. These elements become the basic building blocks for the majority of our consumer and industrial chemicals. In this category, chemicals made from petroleum base are particularly crucial in the operation of technologies used in refrigeration and cooling.

Using petroleum, dichlorodifluoromethane, which became known as Freon, was invented by Thomas Midgley Jr. with co-inventor Charles Kettering to serve as an alternative to the toxic gases that were previously used as refrigerants, such as ammonia, chloromethane, and sulfur dioxide. This breakthrough led to an entire family of related chemicals, with each Freon product designated by a number, including Freon-11 (trichlorofluoromethane) and Freon-12 (dichlorodifluoromethane), which are coolants, and Freon-113 (trichlorotrifluoroethane), which is a cleaning agent. Freons were useful but dangerous from the start: for instance, if its temperature rises higher than 400 degrees Fahrenheit, Freon converts to phosgene gas, commonly known as nerve gas—the agent that gained notoriety in World War I for its sweet smell of cut grass. In this form, the gas caused 90,000 deaths in "the war to end all wars" and 350,000 in World War II (excluding Nazi use of gas chambers), which was discussed above.

Chlorofluorocarbons (organic chemical compounds often used in air conditioning and other purposes) and Freon in particular became one of the first foci of the modern environmental era as well as an initial lesson about the problematic existence we had come to in our new "ecology of oil." Devra Davis writes of ozone and CFCs—what she calls "free radicals"—in this fashion:

> In the lower atmosphere, CFCs are basically inert. . . . But when they float up to the stratosphere, where they are exposed to stronger ultraviolet rays, decomposing molecules of CFCs release atoms of chlorine. Each chlorine atom can destroy tens of thousands of molecules of ozone . . . that serves as a global sun shield.[28]

Well before public discourse over the concept of climate change, the ability of the "free radical" CFCs, such as Freon, to create damage that affected every human was profound.

The use of these chemicals, such as DDT, and their production has revealed that despite great advantages, the creation of petrochemicals often comes with serious environmental costs. In the United States, most of this production has been concentrated in the American South; for instance, the location of the facilities to produce benzene and other petrochemicals is one of the most significant instigators of cases of environmental justice and racism. Regulation and careful environmental monitoring have managed to somewhat limit the impact of such sites; most often, however, the reaction of the industry has been simply to move the production process overseas to a less regulated site.

Although each of these uses of petroleum presents some deleterious outcomes for human health, the production of petrochemicals, which were typically established near necessary sources of massive quantities of petroleum, seriously affected communities all over the world. The primary example of the industry's possible outcomes occurred in Bhopal, India, in December 1984 when a gas leak occurred at the Union Carbide India Limited pesticide plant, killing over thirty-seven hundred residents and injuring over half a million others; however, the twenty-first century is littered with other sites teetering on the brink of similar cataclysm.[29]

CONCLUSION: FROM COLD WAR TO RESOURCE WARS

Cold War stratification held parts of the world in check for decades. Some of the earliest fractures in this bipolar world appeared through the growing importance of energy resources, particularly crude oil, as they transcended the imposed ideological boundaries of nations. For nations possessing crude as well as those needing it, such developments marked a foretaste of a new world order that emphasized access to energy.

This was a global structure initiated by the Cold War superpowers. In 1950, a Jersey Oil Company executive said simply: "It appears that in the future, Mideast crudes . . . may exceed requirements substantially."[30] By 1960, independence movements and decolonization influenced many of the nations of the Middle East. With this increasing autonomy, oil-producing nations sought to rectify the exploitative arrangements by banding together. In 1960, the oil exporting nations joined forces to combat the unfettered influence of international oil companies by establishing the Organization of the Petroleum Exporting Countries (OPEC), which will be discussed further in chapter 7. During the subsequent years, OPEC gained political clout through some activities of its own but also through the fuel dependence of developed nations such as the United States. Between 1948 and 1972, consumption in the United States grew from 5.8 million barrels per day to 16.4. This three-fold increase was surpassed by other parts of the world: Western Europe's use of petroleum increased by sixteen times and Japan by 137 times. Throughout the world, this growth was tied to the automobile: worldwide, automobile ownership rose from 18.9 million in 1949 to 161 million in 1972; the U.S. portion of this growth was significant, from 45 million to 119 million during the same years. New technologies enabled some refiners to increase the yields of gasoline, diesel and jet fuel, and heating oil from a barrel of petroleum, but the needs remained unlike anything the world had ever seen.

Such reliance on fuel forced the U.S. government to consistently question relevant policies. In 1969, the administration of Richard Nixon began debating its quota or limitation on importing oil. In April 1973, Nixon delivered the first-ever presidential address on energy, in which he announced that he would abolish the quota system and thereby thrust the United States forward as a full competitor in the global marketplace for importing crude oil. The new reality, however, was that domestic production could not keep up with the American needs. Clearly, quotas were meant to manage and limit supplies of crude oil in a surplus market, not in the world of shortages that was taking shape. Without the import barriers, after years of producing its own supply of crude, the United States was a full-fledged and very dependent member of the world oil market. At the end of the Cold War in 1990, the global landscape of petroleum had changed in at least one very dramatic way: energy was essential to the development of *any* nation and the United States would need to queue up with all other nations to acquire a consistent supply of crude.

NOTES

1. University of Bristol, "Researchers Venture to the Chernobyl Red Forest." https://phys.org/news/2019-04-venture-chernobyl-red-forest.html. Accessed on January 6, 2022.

2. Rachel Nuwer, "Forests Around Chernobyl Aren't Decaying Properly." https://www.smithsonianmag.com/science-nature/forests-around-chernobyl-arent-decaying-properly-180950075/. Accessed on January 6, 2022.

3. Robert B. Marks, *The Origins of the Modern World: A Global and Ecological Narrative* (Lanham, MD: Rowman & Littlefield, 2002), p. 179.

4. Paul Boyer, *By the Bomb's Early Light* (Chapel Hill: University of North Carolina Press, 1994), pp. 36–37.

5. Thomas Hughes, *American Genesis* (Chicago: University of Chicago Press, 2004), pp. 143–145.

6. Hughes.

7. Boyer, pp. 95–96.

8. Richard Rhodes, *Energy* (New York: Simon & Schuster 2019), p. 284.

9. Boyer, p. 200.

10. John McNeill and Peter Engelke, *Great Acceleration* (London: Belknap Press, 2016), p. 156.

11. Rhodes, p. 284.

12. McNeill and Engelke, p. 7.

13. Rhodes, p. 287.

14. Rhodes, p. 290.

15. Rhodes, pp. 282–283.

16. Marks, p. 179.

17. McNeill and Engelke, p. 7.

18. McNeill and Engelke, p. 312.

19. Alfred Crosby, *Children of the Sun* (New York: Norton, 2006), p. 137.

20. McNeill and Engelke, p. 168.

21. McNeill and Engelke, pp. 312–314.

22. McNeill and Engelke, p. 313.

23. McNeill and Engelke, pp. 163–164.

24. McNeill and Engelke, p. 159.

25. John D. Grace, *Russian Oil Supply* (New York: Oxford University Press, 2005), p. 287.

26. Todor Balabanov and Raimund Dietz, "Eastern and East West Energy Prospects," in *Dismantling the Command Economy in Eastern Europe*, ed. Peter Havlik (Boulder, CO: Westview Press, 1991), pp. 125, 137.

27. Jeff Meikle, *American Plastic* (New Brunswick, NJ: Rutgers University Press, 1995), p. 78.

28. Devra Davis, *When Smoke Ran Like Water* (New York: Basic Books, 2002), p. 250.

29. Apoorva Mandavilli, "The World's Worst Industrial Disaster Is Still Unfolding," *The Atlantic*, July 10, 2018. https://www.theatlantic.com/science/archive/2018/07/the-worlds-worst-industrial-disaster-is-still-unfolding/560726/. Accessed on January 6, 2022.

30. Daniel Yergin, *The Prize* (New York: Free Press, 2008), p. 430.

TRANSITIONING BY THE NUMBERS
High-Energy Existence

The Great Acceleration of energy exchanges that shapes and defines the Anthropocene sees the construction of modern living patterns that integrate and expand systems that predominantly begin with fossil fuels. With these expectations of energy supply in place, energy-intensive living patterns profoundly separate humans living in developed and less developed societies.

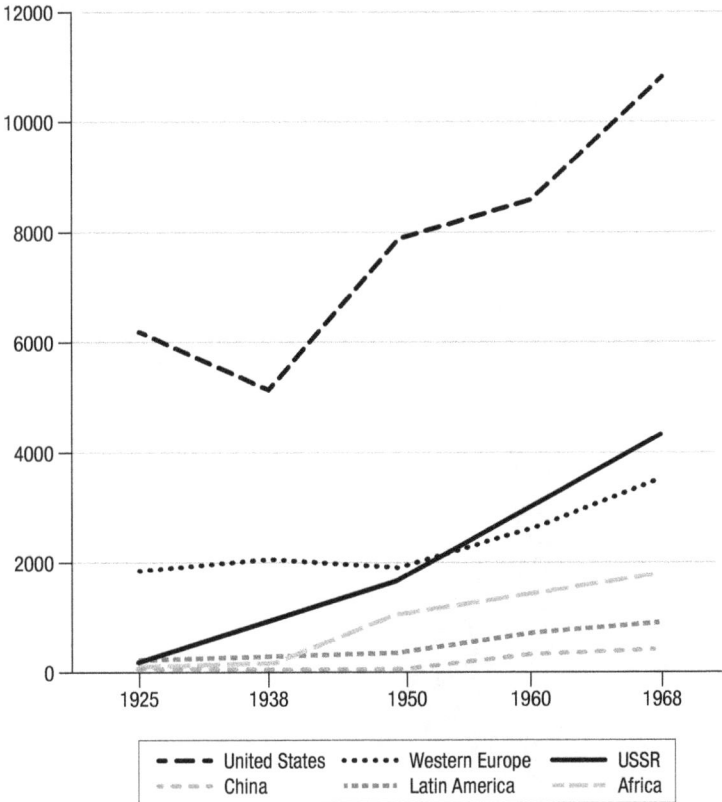

Powering the Great Acceleration, by nation, 1925–1968 (kilograms of coal equivalent per person per year). ASTRID KANDER, PAOLO MALANIMA, AND PAUL WARDE, *POWER TO THE PEOPLE: ENERGY IN EUROPE OVER THE LAST FIVE CENTURIES* (PRINCETON, NJ: PRINCETON UNIVERSITY PRESS, 2014).

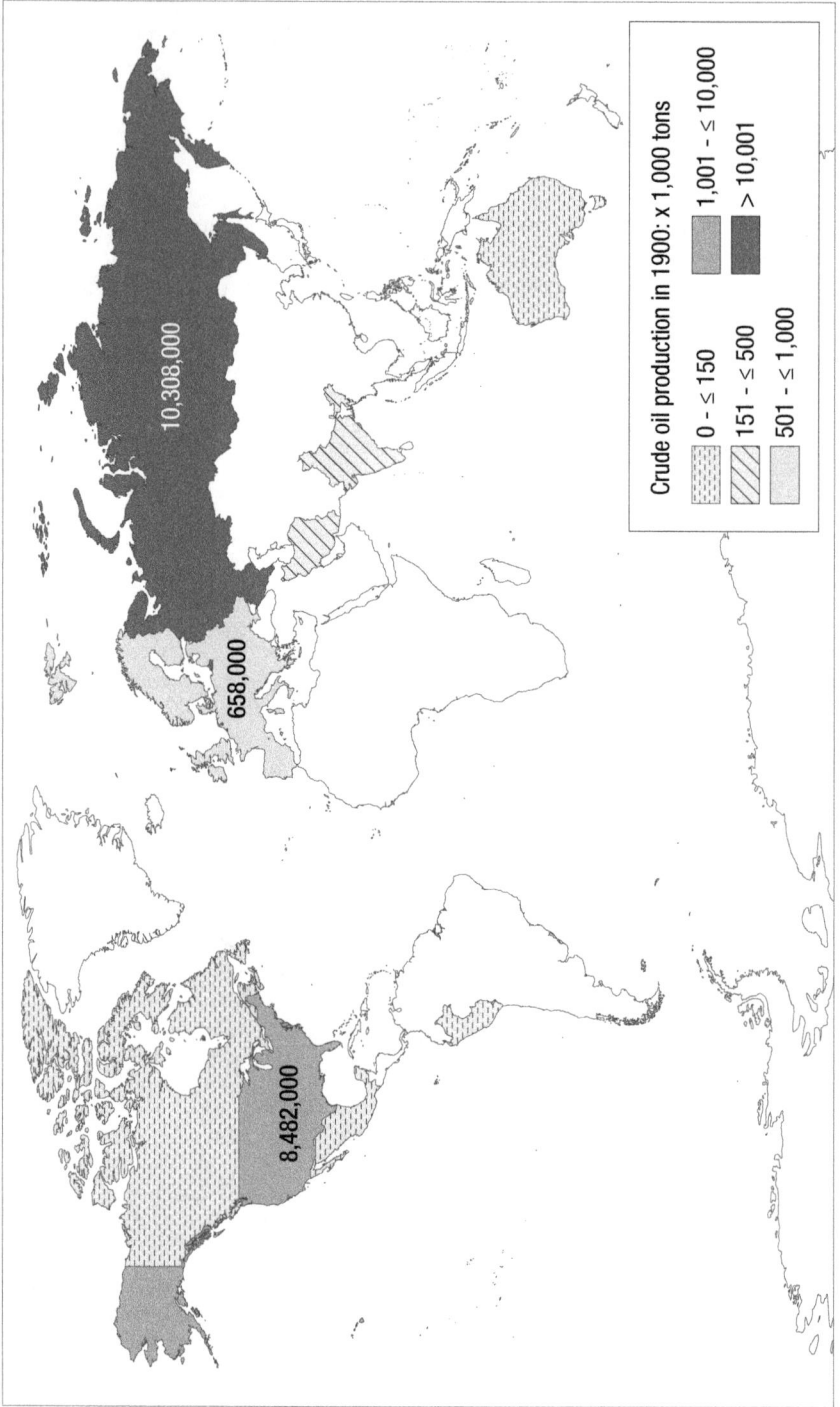

Crude oil production in 1900: x 1,000 tons

0 - ≤ 150	1,001 - ≤ 10,000
151 - ≤ 500	> 10,001
501 - ≤ 1,000	

10,308,000

658,000

8,482,000

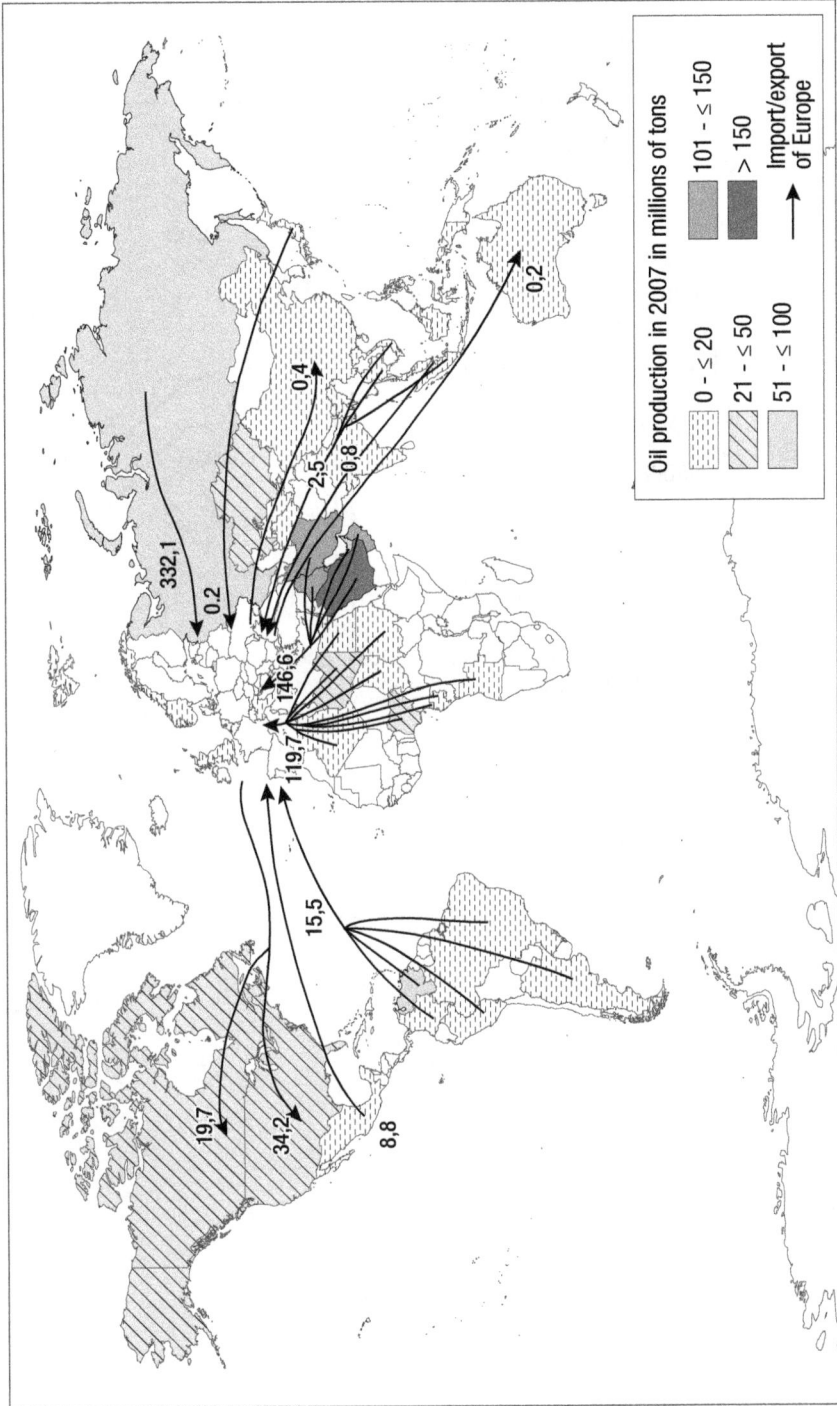

Production of World Oil in 1900 and European Imports of Oil in 2006. ASTRID KANDER, PAOLO MALANIMA, AND PAUL WARDE, *POWER TO THE PEOPLE: ENERGY IN EUROPE OVER THE LAST FIVE CENTURIES* (PRINCETON, NJ: PRINCETON UNIVERSITY PRESS, 2014).

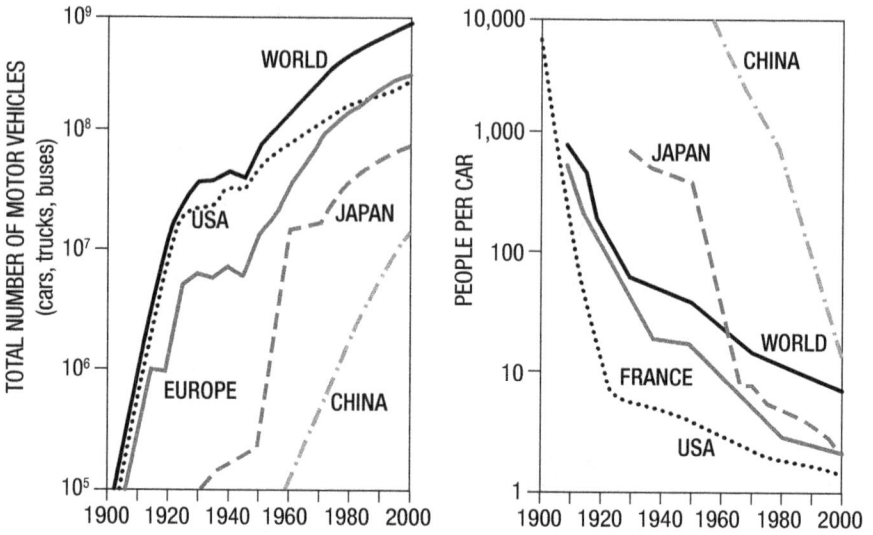

World Vehicle Ownership and by Nation, 1900–2000. VACLAV SMIL, *ENERGY AND CIVILIZATION: A HISTORY* (CAMBRIDGE, MA: MIT PRESS, 2017).

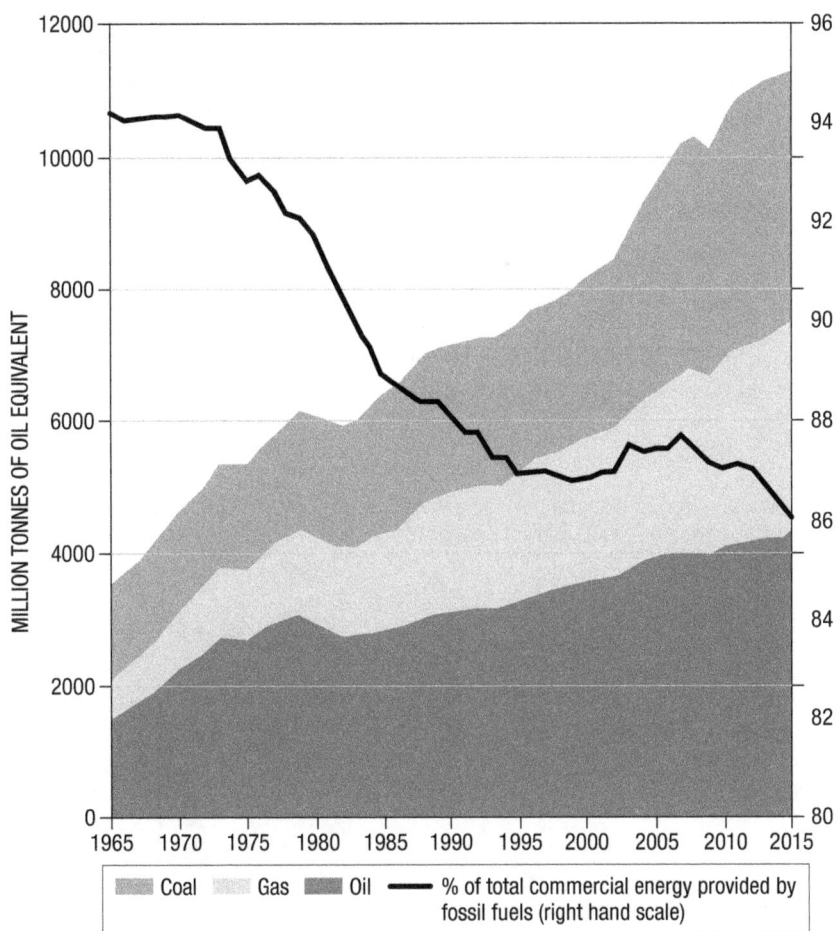

Global Fossil Fuel Consumption and Decline in Commercial Energy, 1965–2015. SIMON PIRANI, *BURNING UP: A GLOBAL HISTORY OF FOSSIL FUEL CONSUMPTION* (LONDON: PLUTO PRESS, 2018).

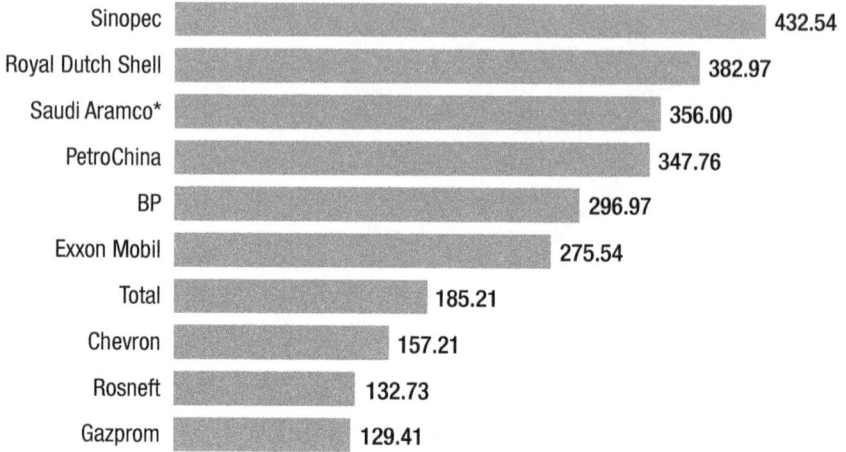

Sinopec	432.54
Royal Dutch Shell	382.97
Saudi Aramco*	356.00
PetroChina	347.76
BP	296.97
Exxon Mobil	275.54
Total	185.21
Chevron	157.21
Rosneft	132.73
Gazprom	129.41

*2018

Largest Oil and Gas Companies in the World in Billion U.S. dollars, 2019. HTTPS:// WWW.STATISTA.COM/CHART/17930/THE-BIGGEST-OIL-AND-GAS-COMPANIES-IN-THE-WORLD/. ACCESSED JANUARY 6, 2022.

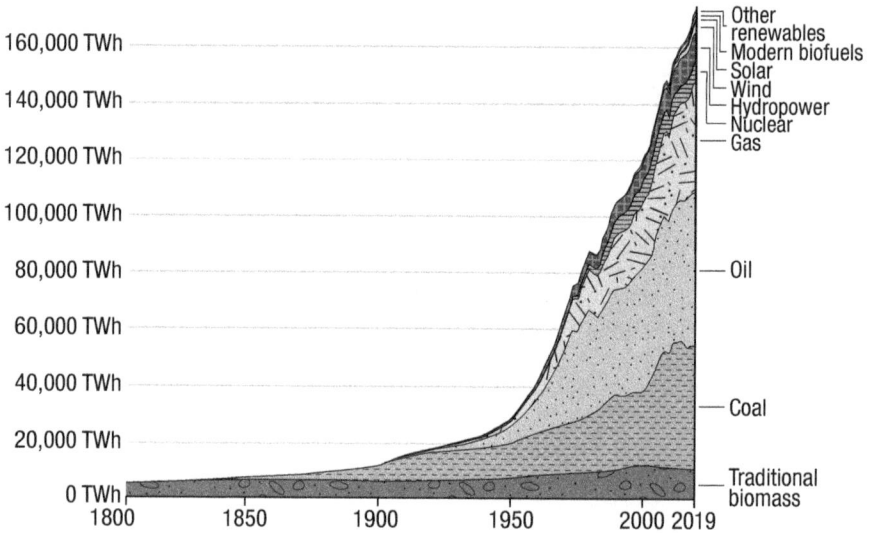

Global Energy Use by Source

Note: Measured in terawatt-hours (TWh), this graphic reveals a consistent growth in the diversity of methods of power generation worldwide as well as the massive expansion of overall supply.

7

The Energy Gap Takes Shape

GROUND: THE HONDA CIVIC AND THE
REALIZATION OF AUTOMOBILE EFFICIENCY

Dodge Royal Monaco, Ford LTD, Buick Riviera, Chrysler New Yorker, Chevy Impala, Cadillac Eldorado—to name only a few of the great American autos that entered the 1970s as status symbols of individual financial success. In just a brief moment, the glamour of the large, heavy, American-manufactured autos—indeed, the pride and symbol of post-1950s consumer culture—became dinosaurs of a previous era. The American roadways of the 1970s and other global markets made way for smaller, more efficient competitors, such as the Japanese-made Honda Civic that arrived in global markets in 1973. Particularly in the 1970s, before American manufacturers could reconsider their designs, this provided a dramatic inroad for globalizing the auto market to include alternative designs. With an eye toward efficiency, consumers in developed nations flocked to Datsun, Toyota, Honda, and the small vehicles made in Japan.

Although Japan was only a minor player in the emergence of consumer autos in the early twentieth century, its vehicles became a market leader when efficiency emerged as a significant consumer consideration during the 1970s. In 1970, Japanese four-wheeled motor vehicle exports reached the 1,090,000 mark, surpassing Italy, the United States, and the United Kingdom (in that order) and making Japan the world's third largest automobile exporter.[1] Why had Japan emphasized smaller, more efficient vehicles from the start?

In short, Japan never had supplies of crude to support inefficient, decadent vehicle design. In Japan, the first world oil crisis of 1973 spiked the price of gasoline by 217 percent. Conserving petroleum became a national policy and Japanese manufacturers also implemented the world's strictest exhaust emissions regulations. Japanese manufacturers became particularly adept at squeezing maximum mileage out of each drop of gasoline. Lighter vehicle weight is one factor contributing to increased fuel efficiency. An obvious way to decrease weight is to reduce vehicle size, but this was not much of an option for Japanese manufacturers who had been producing small cars for decades, unlike U.S. and European manufacturers. One weight-reducing solution was the adoption of the front-wheel-drive system; another was the use of lighter parts and materials, such as high-tension steel sheeting, aluminum, and plastics. Additional measures to increase fuel efficiency were the introduction of electronically controlled fuel injection systems, reduced air resistance, and the use of new-technology materials such as fiber-reinforced metals, plastics, and ceramics.

Between 1975 and 1985, the competition between small passenger cars and popularly priced mini vehicles was the driving force behind new product development and market expansion, particularly in Japan. Small passenger car sales took the lead in the aftermath of the oil crisis, because new models were developed that complied with the government's strict exhaust emissions regulations and were also very competitively priced. In 1976, automobile production became the leading domestic manufacturing industry for Japan. By the end of the decade, on to American roads came the Ford Pinto, Chevrolet Vega, AMC Gremlin, and the imported Plymouth Cricket, Dodge Colt, and Volkswagen Beetle, all of which followed the downsized Japanese design of the Honda Civic, Toyota Celica, and Datsun/Nissan Fairlady, Bluebird, and Sunny 1200.

Throughout the globe, energy became a primary dimension of national development after 1970—not just for the Cold War superpowers. In scale, the capabilities of modern humans dwarfed all seen previously. For instance, damming the Yangtze River had been suggested as early as 1919 by nationalist Sun Yat-sen;

however, it was not until 1989 that the national government in China chose to proceed with the world's largest water project. Although the project always had great potential for irrigation, flood control, and power generation, the social and political uprisings that climaxed in 1989 spotlighted the project's ability to symbolize that the Communist Party remained in charge of the new China that was emerging. The Three Gorges project now stands as the world's largest, with a reservoir roughly the length of Lake Superior and its construction resulting in the relocation of approximately two million rural Chinese.

Throughout the twentieth century, China's hydropower efforts have been enormous. More than half of the dams exceeding fifteen meters in height in the entire world are in China, almost all of them built since 1949. Historian Kenneth Pomeranz writes that in the twenty-first century, "China still has a great deal of untapped hydropower potential—the most in the world, in fact—and, unlike those of some richer countries, its government does not seem to be having many environmentalist qualms about big dams." Without natural gas reserves, China also has few clean energy options.

While emphasizing energy centrality to its national planning, China today, in Pomeranz's view, resembles the Cold War powers of the twentieth century in its love of systems. In China, he writes,

> a group of technocrats favors very large-scale engineering projects that pose enormous environmental risks . . . wield considerable power: a remarkable percentage of China's top leaders in recent decades have been engineers. The lack of constitutional limits on central power in China makes the realization of these projects all the more likely. Not only is the ability of affected citizens (such as those scheduled for relocation) and other dissenters (such as environmentalists) to protest severely constrained, but even local and provincial governments lack a firm legal base from which to object to national appropriation of "their" resources.[2]

China, of course, pursues the Three Gorges project while simultaneously emerging as the world's leader in the manufacture of solar and other renewable energy sources. The common theme of these projects is the shared priority of autonomous, reliable power development.

In recent decades, the emerging reality of energy centrality took a variety of forms around the world. Existing nations, ranging from China to Sweden and Norway to Brazil, created national energy agendas, while emerging states sought the most effective route to catch up to the "haves." For a globe in transition after the Cold War, access to energy became a common agenda item among nations, particularly among nations that emerged from the era of colonialism that extended roots back to the 1500s. Independence movements had begun after World War II and continued during the Cold War. Both the United States and the Soviet Union had strongly anticolonial ideologies and were opposed to

European states maintaining their colonies. The United States wanted to end colonialism on a human rights perspective and had an interest in competing in those markets for raw materials that were controlled by colonial authorities. Soviet interest in ending colonialism derived more from the possibility of social-ist transformations that might serve as a step toward a communist world. In many parts of the world, decolonization joined with Cold War interests to create an active competition for development and influence.

The movement for independence allowed for an emphasis on modernization, urbanization, and commodities that particularly coheres with the trajectory of the narrative of this book. In particular, energy helped to form a basic tenet of the new world order; access to energy resources marked a direct line to standing on the world stage. Particularly during the 1980s, writes economic historian Simon Pirani, "a blizzard of motorization hit developing world cities, encouraged by international agencies and governments who ignored everything urban plan-ners had learned about the damage done by car-based cities in the rich world." For instance, by 1990 Mexico City had an estimated four million cars—forty times as many as in 1950.[3] In terms of electricity, the trends were similar: in 1950, one-tenth of the world's fossil fuels were used to produce electricity; by 2011, more than one-third of the fossil fuels generated electricity. Pirani writes:

> In 1970, most (51 percent) of the world's 3.7 billion population, including three quarters of the people outside the rich world, had no electricity. Two decades later, in 1990, about 3.2 billion of the world's 5.3 billion population had electricity, in-cluding just under half of those living outside the rich world. Another two decades later, in 2013, 6 billion of the world's 7.2 billion people had access, including more than three-quarters of people outside the rich world.[4]

The era of globalization tied all nations into shared markets and suggested some common goals leading toward economic and political stability that placed at the fore smooth, available energy supplies. The importance of energy almost appeared as a secret that had emerged among a few powerful nations at the start of the twentieth century, became the unstated route to economic dominance during the Cold War, and now forged a new structure to the post–Cold War world order.

In this post-1990 world, the clear importance of energy resources combined with dwindling or transitioning supplies to place new importance on those na-tions that had access to power and those that did not. This spirit drove China's Three Gorges project as well as many more emphasizing an assortment of energy sources. As this larger pattern played out in the postcolonial era, however, the flash point of economic power and conflict was often one transnational energy source in particular: crude oil.

CURRENT: Decolonization and the Manipulation of Energy Scarcity

In the twenty-first century, the value of crude is well known and nations possessing it openly leverage its value in order to benefit themselves. This point in the geopolitical structure of the world differs significantly from that of a century earlier when global corporations and the nations behind them bullied and dominated nations possessing petroleum who were largely unaware of its worth. The key to this transition was the 1970s when automobile lines at gas stations in the United States and Europe functioned as one indicator of massive changes in global affairs after World War II.

"Decolonization" refers to this era when many additional nations, loosed from colonial authority, became autonomous, responsible for their own development and governance. Although the Cold War added a new version of quasi colonial authority as the American and Soviet diplomats vied to spread their ideology and to squelch that of their opposing superpower, overall, nations in Africa and particularly the Middle East could begin to pursue their own futures. Even if they lacked the internal political infrastructure to do so, they each sought to expand what power they did possess. In this new, unfettered political environment, the use and management of every resource took on strategic importance and, therefore, it follows that the administration of the world's most sought after commodity reflected these changes most acutely. Simply, when petroleum supplies stuttered during the 1970s, there was no safety net to catch consumers in "have" nations, particularly American and European—there existed no federal method for offsetting the temporary glitches in supply.

From the stranglehold of Western powers and the large petroleum corporations that dominate them, oil grew into a tradable, ultravolatile commodity. Yergin writes that this new era in world oil demonstrated that "oil was now clearly too important to be left to the oil men."[5] As political leaders in each oil nation assessed how best to leverage power for their nation from their supply of crude, it took little time for them to also realize the merit of joining forces with similarly endowed nations. Joining forces would allow oil-producing nations to control supply and prices and, finally, to gain a competitive advantage in negotiations with transnational corporations.

This was the essential goal in September 1960 when nations formed the Organization of the Petroleum Exporting Countries (OPEC) in Iraq. Its formation was precipitated by changes in the oil market after World War II and driven by the new status of many of these less developed nations. Lacking exploration skills, production technology, refining capacity, and distribution networks, oil-producing countries were unable to challenge the dominance of the oil companies prior to World War II.[6] OPEC allowed oil-producing nations to hold sway against powerful oil corporations that had dominated them in the previous era of oil exploration. It was one of the first large-scale, international political groups framed around a single resource—a cartel. OPEC's founding members in 1960 were Iran, Iraq, Kuwait, Saudi Arabia, and Venezuela. Eight other countries joined later: Qatar (1961), Indonesia (1962), Libya (1962), United Arab Emirates (1967), Algeria (1969), Nigeria (1971), Ecuador (1973), and Gabon (1975) (Ecuador and Gabon withdrew from the organization in 1992 and 1994, respectively).

Across differences of location, climate, religion, and political structure, these nations had the common concern of oil. To varying degrees, though, they also shared small size and a lack of political influence on the world scene. Bound together, the countries of OPEC had an obvious purpose: to manage supplies of crude on the market in order to maintain high prices and, thereby, leverage the profits of member nations. In short, they sought to exploit and leverage the culture of petroleum that pumped through nations such as the United States and to exploit the emerging geopolitical situation: the increasing scarcity of petroleum supplies in the face of its necessity in developed nations.

It seems ironic today to talk about oil producers—whether corporations or nations—needing to manipulate markets in order to keep the price of petroleum profitable; yet, as we noted above, major oil companies colluded among themselves and with national powers through tools such as colonialism from the 1920s to the 1960s to prevent prices (and profits) from falling. As these corporations' influence waned, other methods were employed. One of the most significant difficulties was that as prices fell, domestic producers simply could no longer compete. Moreover, during the 1950s the Eisenhower administration concluded (as the Japanese had prior to World War II) that dependence on foreign oil placed U.S. national security in jeopardy. The United States responded by implementing import quotas, which were intended to keep domestic prices artificially high and to represent a net transfer of wealth from American oil consumers to American oil producers. By 1970, the world price of oil was $1.30 per barrel and the domestic price of oil was $3.18.[7]

OPEC AND THE "OIL WEAPON"

OPEC's ability to manipulate prices did not fully become a reality until Egyptian leader Anwar Sadat urged members to "unsheathe the oil weapon" in early 1973.

The primary rationale for this action was politics. Israel's military aggression outraged its Arab neighbors throughout the late 1960s. Israel's attack on Egypt in 1967 resulted in an earlier embargo, which proved unsuccessful because of oversupply of crude on the world market. In October 1973, U.S. president Richard Nixon agreed to provide more military jets to Israel after a surprise attack on the nation by Egypt and Syria. On October 19, the Arab states in OPEC (Organization of Arab Petroleum Exporting Countries; OAPEC) elected to cut off oil exports to the United States and to the Netherlands.

In petroleum circles, the embargo is often referred to as the "first oil shock." As such, it combines new market features of the early 1970s: first, production restraints that were ultimately supplemented by an additional 5 percent cutback each month; and, second, a total ban on oil exports to the United States and the Netherlands and eventually also to Portugal, South Africa, and Rhodesia. Factoring in for production increases elsewhere, the net loss of supplies in December 1973 was 4.4 million barrels per day, which accounts for approximately 9 percent of the total oil available previously.[8] Although these numbers told of a genuine shortfall in the overall supply, the fickle petroleum market accentuated the embargo's importance by inserting a good bit of uncertainty and panic. It was the American consumers who felt the impact most, because they had grown so completely accustomed to a culture defined by petroleum abundance.

In order to provide oil to consumers, brokers began bidding for existing stores of petroleum. In November 1973, per barrel prices had risen from around $5 to more than $16. Foreshadowing patterns of the twenty-first century, consuming nations bid against each other in order to ensure sufficient petroleum supplies. For American consumers, retail gasoline prices spiked by more than 40 percent. Although high costs were extremely disconcerting, scarcity also took the form of temporary outages of supply. American consumers previously content to drive their cars until gas gauges neared empty now lined up for a few gallon ration whenever it was available. One journalist described the scene near New York City in this fashion:

> Anxious motorists overwhelmed gasoline stations in the metropolitan [New York] area yesterday, with many stations running out of supplies early in the day, while dealers hoped incoming deliveries under February allocations would restore calm by mid-week.
>
> In Brooklyn, Murray Cohen, an owner of the AYS Service Station at Avenue Z and East 17th Street, said he had imposed a $3 maximum for each car's purchases, only to find that most people needed only 75 cents' worth to fill up. One man, he said, waited in line for an hour and could use only 35 cents' worth.
>
> In Washington, William E. Simon, director of the Federal Energy Office, who had asked drivers not to buy more than 10 gallons at a time, yesterday issued an appeal to them to stay away from stations unless they bought at least $3 worth. . . . "Panic buying isn't helping the situation."[9]

Many American states implemented odd/even day gas purchasing based on the car's license plate number.

As the nation most defined by the new era of petroleum consumption, the United States had the rudest awakening during this period of false scarcity. Intermittently, U.S. motorists throughout 1973–1974 needed to wait in line for one to two hours or more—often, ironically, with their engines running the entire time. In other regions, the worst harbinger became signs that read: "Sorry, No Gas Today." Expressway speeds were cut from sixty–seventy miles per hour to fifty. Some tolls were suspended for drivers who carpooled in urban areas. Even if communities did not implement rationing plans, the American culture of petroleum was altered (at least temporarily) by plans being leaked to the public. For instance, in the New York City region the federal energy office estimated that residents eighteen years of age and older could expect to receive books of vouchers for thirty-seven gallons per month.[10]

By the end of 1973, in fact, gas lines were plentiful throughout the United States and Europe—the "haves." Supplies of petroleum were least disturbed on the West Coast, but by February even California had adopted odd/even day rationing. Gas station operators were subjected to mistreatment, violence, and even death threats and attacks. Drivers also reacted with venom to other drivers attempting to cut into gas lines. At the root of such anger, of course, was the cruel reality that the events of humans' everyday lives—kids going to school, adults going to work or shopping, goods moving in every direction, and even cutting grass—might be constrained, humans in developed societies such as the United States finding their choices limited. Nothing could seem more discordant to the ideals of expansive consumption. For these reasons, the implications of the 1970s crisis were diverse and transformative, particularly for the nation most dependent on oil imports.

CONFRONTING AMERICAN CONSUMPTION

While the embargo had economic implications, it had begun as a political act by OAPEC nations and, therefore, the American president Richard Nixon dealt with it on a variety of fronts, including international political negotiation. These negotiations were based on the emergent geopolitical organization of the world, even if they actually had little to do with the trade of black gold. Negotiations heated up on a number of fronts, including between Israel and its Arab neighbors; between the United States and its allies; and between the oil-consuming nations and the Arab oil exporters. There was a new urgency to the interconnection of trade networks, and establishing a discourse that demonstrated mutual respect became a priority for all concerned. Convincing Arab exporters that negotiations would not begin while the embargo was still in effect, the Nixon administration leveraged the restoration of production in March 1974. However, staring into the face of petroleum scarcity stirred many American politicians to consider new options for reforming American patterns of consumption. For a brief time, some American consumers were also willing to admit that their culture of mass consumption might be unsustainable.

Behind the scenes, the embargo and supply difficulties considerably shifted internal relations in the Middle East. In *Oil Kings*, historian Andrew Scott Cooper demonstrates how the Shah of Iran worked separately with President Richard Nixon to confront OPEC's power structure and production limits. "With a vast supply of petrodollars and U.S. weapons pouring in," writes Cooper, "there seemed to be nothing to stop the empire of Iran and its Shahanshah from dominating not only the Persian Gulf and the land bridges into Central Asia, but even . . . down along Africa's east coast."[11] Trading arms and flexing quotas in the early 1970s, the U.S.–Iranian relationship actually destabilized the Shah's power in the region and in his own nation. When, after 1975, the United States increased oil imports from Saudi Arabia, the Shah's standing was undercut further, and he was left largely unprotected.

The problem of petroleum imports became even more complex when, at the end of the decade, just as some Americans might have begun to think problems with the petroleum supply were a thing of the past, relations with Iran took a turn for the worse. When Iranians took Americans hostage in 1979, U.S. president Jimmy Carter placed an embargo on the importation of Iranian oil into the United States and froze Iranian assets. The "second oil shock" followed, only to be exacerbated in 1980 when the Iran-Iraq War abruptly removed almost four million daily barrels of oil from the world market—15 percent of total OPEC output and 8 percent of free world demand.

CONSIDERING OPTIONS WITHIN THE HIGH-ENERGY EXISTENCE

In the United States and many developed nations, for many activists, politicians, and planners, the 1970s seemed to emerge as a hinge that would lead them away from their dangerous dependence on petroleum now largely imported from other nations. For many consumers, things had clearly changed in the consumptive nature of American life. In particular, the American idea of energy use—in its broadest sense—was brought under new scrutiny. This impact could be seen most clearly in the oval office of U.S. president Jimmy Carter, a trained nuclear engineer. Carter was moved to consider deeply the ways that American society insatiably consumed energy. He reflected on revolutionary new ideas such as that put forward by economist Amory Lovins in a 1976 *Foreign Affairs* article entitled "Soft Energy Paths." In his subsequent book, Lovins contrasted the "hard energy path," as forecast at that time by most electrical utilities, and the "soft energy path," as advocated by Lovins and other utility critics. He writes:

> The energy problem, according to conventional wisdom, is how to increase energy supplies . . . to meet projected demands. . . . But how much energy we use to accomplish our social goals could instead be considered a measure less of our success than of our failure. . . . [A] soft [energy] path simultaneously offers jobs for the unemployed, capital for businesspeople, environmental protection for conservationists, enhanced national security for the military, opportunities for small business to innovate and for big business to recycle itself, exciting technologies for the secular, a rebirth of spiritual values for the religious, traditional virtues for the old, radical reforms for the young, world order and equity for globalists, energy independence for isolationists. . . . Thus, though present policy is consistent with the perceived short-term interests of a few powerful institutions, a soft path is consistent with far more strands of convergent social change at the grass roots.[12]

In addition to promoting coal and nuclear power, Carter took the ethic of energy conservation directly to the American people.[13] His administration would be remembered for events such as the Iranian hostage crisis; however, when he controlled the agenda he steered American discourse to issues of energy. In a 1977 speech, Carter urged the nation:

> Tonight I want to have an unpleasant talk with you about a problem unprecedented in our history. With the exception of preventing war, this is the greatest challenge our country will face during our lifetimes. The energy crisis has not yet overwhelmed us, but it will if we do not act quickly.
>
> It is a problem we will not solve in the next few years, and it is likely to get progressively worse through the rest of this century.
>
> We must not be selfish or timid if we hope to have a decent world for our children and grandchildren.
>
> We simply must balance our demand for energy with our rapidly shrinking resources. By acting now, we can control our future instead of letting the future control us. . . .
>
> Our decision about energy will test the character of the American people and the ability of the President and the Congress to govern. This difficult effort will be the "moral equivalent of war"—except that we will be uniting our efforts to build and not destroy.[14]

In a risky political move, Carter attempted to steer Americans down a path less trodden—in fact, a path requiring severe difficulty and radical social and cultural transition. The postwar ideals of "Futurama" and "muscle" cars, he argued, needed to prioritize resource management inspired by the concept of restraint and conservation.

It was a lonely argument to make in the United States during the late 1970s when the vast majority of Americans knew little of environmental perspectives. The ethic that he described focused on the ethic behind conspicuous consumption that lay at the foundation of the American ecology of oil. Although he offered a clear vision of our limited future based on increasingly scarce, extracted energy resources, by the 1980s many Americans were returning to business as usual—or worse. In this reaction to the 1970s oil crisis, however, Americans were an exception among developed nations. In many other nations, the hinge effect of the 1970s brought an intellectual end to any illusions of conspicuous consumption organized around the wasteful use of petroleum. It marked a paradigm shift toward new ideas about energy that was organized most by renewable sources.

In some other developed nations, centralized authority—including forms of socialism—became a tool for more quickly implementing the lessons of the 1970s in the form of taxes, mass transit development, and diversification of energy supplies. The reaction to the 1970s hinge created a clear gap within the developed nations, separating those that institutionalized the transition of the 1970s (particularly the European Union) and the one that regressed (the United States)—creating a new variation on the idea of "haves and have-nots" within the developed nations. Within the next few decades, overreliance on imported crude would define a new era of international relations and even rationales for waging war.

Events of the 1970s shook the basic structure of the existing energy culture with enough force that radical changes were given new attention. Governments throughout the developed world quickly came to the realization that reliance on fossil fuels compromised their economic and national security and they sought a variety of methods for assuring future energy supplies, short and long term. In short, OPEC's action set about dramatically different reactions among nations and, therefore, we see very distinct potential energy paths—offsetting currents—that will be discussed below. Regardless of their distinction from each other, though, each of these lines of action derived from a basic realization of energy's essentialness—which was also the realization at the root of OPEC's action.

CURRENT: Greening the Energy Imperative with Alternatives

Energy shortages in the 1970s also spurred consuming nations to look more seriously at alternative sources of power. Many of these sustainable approaches to power generation, ironically, reached back to the biological old regime. Although fossil fuels often remained a much cheaper source of power, developed governments clearly began to explore methods of using government subsidies to stimulate the use of alternative energy sources for generating electricity. Two technologies whose growth in popularity after the 1970s significantly altered global energy patterns are solar and wind.

The new emphasis on solar energy technology after the 1970s focused on harvesting it in four different ways: converted directly into electricity using photovoltaic panels; collected and used to heat water or air with the use of a solar thermal collector; solar thermal electricity generated by concentrating sunlight to boil water that can then be used to generate electricity with a steam turbine; and, lastly, passive solar energy integrated into the design of buildings to maximize the amount of sunlight shining through windows to passively heat the building during winter. Regardless of the type of system, any given solar panel or collector will produce more energy when it is used in a more sunny location and, therefore, installations emphasized desirable climates. All four methods of collecting solar energy are pollution free, emit no CO_2, and in most cases use no fresh water.

During the late twentieth century, a focus of the solar industry was to make photovoltaic (PV) panels more affordable. These panels can be installed anywhere that receives sunlight. When installed for electric consumers, PV panels will offset the amount of electricity that must be purchased from the utility, and it is even possible to reduce the annual electric bill to zero except for a monthly customer charge. Photovoltaic installations currently run about $8 per watt for an installed system. At this rate, they are not cost-effective except for off-grid applications. However, in many locations, local, state, and federal incentives are available to reduce the cost of a solar installation. In some locations, these incentives will reduce the cost to as little as $2 per watt. With these large incentives, solar photovoltaic can actual provide electricity at less than 10 cents per kWh making solar cheaper than the utility company.

Having powered the great Age of Sail, wind power also got new attention after 1970. Wind energy is created when the Sun shines on various parts of the Earth and the air in those areas is warmed and expanded. This expanding air is what we call wind and, just like in the early phases of industrialization, a turbine is a device that transforms the kinetic energy of this moving air into electrical energy. Modern wind energy is used only to generate electricity and is limited to only the times when the wind blows. There are methods to store energy from windy times so that the energy can be used when the wind is not blowing, but this is rarely done due to the added cost. Wind turbines can be installed at less suitable locations but will yield a smaller production and the electricity that is produced will cost more. In addition, wind turbines can also be erected in locations where the land use is shared with farming, grazing, or even commercial and industrial uses.

Since wind energy has become cost competitive to current fossil fuel technology (coal and natural gas), there has been a surge in the number of wind farms that have been constructed globally. Wind energy produces no CO_2, no air pollution, and no water pollution. Wind turbines do produce some noise pollution, which may be a concern to adjacent homeowners, and they are said to be an eyesore to some while beautiful to others. But there is much conflicting information about the severity of this noise pollution, the aesthetics of the wind turbine, and the effect on adjacent property values. Wind turbines are known to kill bats, although the reason for this has yet to be found.

Clearly, these technologies resulted in a change in the global energy marketplace after 1970. Additionally, the swift improvement in wind and solar technology prodded nations to experiment with how subsidies and regulations might be used to improve their competitiveness with fossil fuels.

CONDUIT: The Strategic Petroleum Reserve Introduces the Era of Hoarding

Possibly the most obvious response to the petroleum scarcity of the 1970s among a few developed nations was the creation of a tool for stockpiling petroleum supplies—for feigning control over the increasing scarcity. Conceived in the 1970s, the U.S. Strategic Petroleum Reserve (SPR) is the world's largest supply of emergency crude oil. Although Americans of the late twentieth century did not necessarily prioritize conserving fuel, as a matter of policy, SPR demonstrates that a clear lesson had been learned among the nation's strategic planners. From the American perspective, this lesson was best kept from the public. Particularly in relation to SPR, it was to the American advantage not to alert other nations to the drastic measures on which it was about to embark.[15]

Faced with the obvious reality that it required petroleum acquired from elsewhere, the United States made its acquisition a matter of national security while largely hiding the actual logic from American consumers. There was as yet no national discussion of "peak oil," although the creation of SPR was largely an admission of petroleum's finite supply. As such a commodity, petroleum should be hoarded in times of peace and safety so that the nation is best prepared for the scarcity that might arrive when the equilibrium is disturbed for some reason.

Although such a reserve had been considered since the 1940s, the embargo by OPEC in 1973–1974 demonstrated the need for American leaders to possess a reserve in order to offset disruptions in supply, whether caused by political or natural occurrences. President Ford signed the Energy Policy and Conservation Act (EPCA) on December 22, 1975, which declared it to be U.S. policy to establish a reserve of up to one billion barrels of petroleum. The Gulf of Mexico region offered easy access to petroleum shipping and refinery lanes as well as the necessary geological infrastructure: underground salt domes. Brought up to earth's surface in Saudi Arabia or elsewhere, the crude oil was then pumped back into the ground beneath American soil.

The domes had been selected over a few earlier alternatives that were discussed, including a flotilla of tankers and large rubber bags in aboveground locations. Once the caverns were selected, they were readied through a technique called "solution mining," in which water is pushed through the domes and then sucked out until a significant hole underground has been hollowed. Through salt engineering, though, the caverns become water-tight (or oil-tight). The salt wraps itself around the oil like plastic, so the caverns don't leak. During the next twenty years, the federal government spent $37 billion to construct and fill the SPR.

In April 1977, the government acquired several existing salt caverns to serve as the first storage sites (it estimated that five hundred such caverns existed). Construction of the first surface facilities began in June 1977, and in July administrators began to fill them with oil. Although filling continued over the next decades, the public truly only heard of the SPR when a president considered allowing a withdrawal under the authority of the EPCA. In the event of an energy emergency, SPR oil would be distributed by competitive sale. The SPR has been used under these circumstances only twice (during Operation Desert Storm in 1991 and after Hurricane Katrina in 2005).

Today, the U.S. SPR has grown to approximately seven hundred million barrels, and plans are in place to increase it to one and a half billion barrels. Its proponents argue that it is a significant deterrent to oil import cutoffs and a key tool of foreign policy. They argue that in an era of declining petroleum production, the SPR has allowed the United States to overcome its "energy impotence." Would the SPR be as effective if each nation had its own, though? We may need to find out: by 2006, nations declaring some version of their own SPR included each nation in the European Union (this was a requirement of the directive establishing the twenty-seven-nation union), China, Israel, Jordan, Singapore, South Korea, Taiwan, Thailand, Japan, and South Africa; and nations developing reserves include India, Russia, Iran, Australia, New Zealand, and the Philippines.

According to the U.S. Energy Information Administration, approximately 4.1 billion barrels of oil are held in strategic reserves, of which 1.4 billion is government controlled. The remainder is held by private industry. Currently, the U.S. reserve is the world's largest and is contained at two sites in Texas (Bryan Mound, located near Freeport, and Big Hill, near Winnie), two sites in Louisiana (West Hackberry, near Lake Charles, and Bayou Choctaw, near Baton Rouge), and a final site being added at Richton, Mississippi. Although this development is an unbelievably profitable concept to oil producers, it represents the competitive marketplace created by true scarcity. This is not the scene of false scarcity such as that of the 1973 embargo; instead, this scarcity, derived from the concept of "peak oil," brings with it an air of finality. In this revolutionary new logic, as the scientific reality of petroleum supplies comes in line with the culture of consumption, each user races to be the first to acquire the scarce resource on which its society depends.

TRANSPORTATION CREATES THE NETWORK OF GLOBALIZATION

Simultaneous with these efforts to overtly shift energy use patterns after the 1970s, it must also be recognized that the period of decolonization and the end of the Cold War led to new trends of globalization that were often made possible by the expansion or continuation of fossil fuel use, particularly to power transportation infrastructure that increased humans' ability to move about the globe. Individual transportation was the most obvious revolution of human movement related to the high-energy existence; however, this was only a fraction of the change wrought for human ideas of time and space during the twentieth century. On a larger scale, transportation networks powered by petroleum fuels transformed societies through a trade network of roads used for trucking and innovations enabling air travel. Particularly in the case of air travel, human societies expanded nationally and internationally through a system of connections growing from petroleum products. In each case, initial innovations sparked the gradual construction of complex networks that brought humans access over borders and spatial limits previously unmanageable or impenetrable.

On the ground, long-distance trucking presented a flexibility that railroads could not. Predominantly, systems for lorries or trucks proved cost-effective by seizing the petroleum by-product of diesel fuel. Rudolf Diesel developed this fuel and engine design in the late 1870s in hopes of overcoming the inefficiency of other early power systems. Original methods of powering engines applied just 10 percent of innate power supplies to actually moving the vehicle; the diesel compressed air to assist the fuel in raising the engine's temperature and allowed more of the fuel to go toward moving the vehicle. In Diesel's design, fuel is injected into the piston chamber with air, creating an immediate explosion that forces the piston down. Additionally, the fuel needed was different: from the start, diesel fuel allowed flexibility and gasoline could be diluted and mixed with various natural, vegetable oils. When he released his first engine for practical use in 1897, the diesel engine ran at approximately 75 percent efficiency. It used biodiesel made from peanut oil, which Diesel selected in hopes that it would spur smaller, local industries to supplement the gasoline imported from elsewhere. Stationary diesel engines became popular in industry and shipping by 1900; however, the size and weight of the design continued to limit its usefulness.

These limitations led entrepreneurs to tap diesel engines for different types of transportation tasks, particularly trucking. Diesel-powered lorries hit the road in Germany in the mid-1920s, and Mercedes-Benz released the first automobile with a diesel engine in 1936. As diesels became more popular for trucking, particularly in the United States, oil refiners diminished the use of biofuels and created diesel fuel entirely from fossil fuel residues. Largely due to industry preference, biodiesel receded from major markets for approximately a century. The entirely petroleum-based diesel fuel became the primary fuel for American trucking and industry during the twentieth century. Globally, diesel fuel enabled trucking to assist trade in less developed nations as well. In less developed nations, imitations of road development certainly capped trucking's impact, but it did not diminish it entirely. Clearly, the implications of long-distance trucking would be felt most in the expansive United States as it began to organize its commercial future around this flexible mode of transportation.

The use of trucks for more than local deliveries expanded after World War I, during which the United States used approximately six hundred thousand trucks. In 1935, the Motor Carrier Act put the U.S. federal government squarely behind the future of interstate trucking by expanding the purview of the Interstate Commerce Commission beyond regulating rail and water transportation to include regulating trucking companies involved in interstate commerce. Although there were a variety of rationales for the expansion of interstate highways in the 1950s, their single greatest impact was on the long-haul trucking industry. The interstate highways allowed trucking to expand as an industry in a way seen in no other nation. A culture of trucking took shape around the network of truck stops that became essential to drivers remaining on the road for days and weeks at a time. Access to diesel fuel was now needed in remote portions of the United States through which interstates passed and the truck stop became the essential conduit.

In particular, American business wove the trucking network deeply into the nation's commerce through the rest of the twentieth century. In the American countryside, writes historian Shane Hamilton, the expansion of long-haul trucking came "as industrialized agriculture made the practice of farming increasingly peripheral to the economic and social lives of most rural people." Trucking, essentially, facilitated this transition by tying together more disparate operations and offering drivers a consistent occupation in rural regions.[16] Hamilton continues: "Trucks increasingly replaced trains as the transportation mode of choice for farmers and food processors and retailers—not because trucking was inherently 'better' or less costly than shipping by rail, but because trucks provided the technological flexibility required for the new distribution methods."[17] What began as a success for political action in road building and subsidy emerged as a system in its own right by the end of the twentieth century. Mastery of such systems of distribution allowed retail innovations that culminated by the 1980s in Walmart. From truck stops to frozen foods, many innovations were required to achieve the Walmart world of efficient distribution; however, at its foundation was the diesel-powered long-distance trucking industry.

Air travel required much more complex innovations, each of which relied on internal combustion engines powered by petroleum products. Although early flying machines were powered by a variety of petroleum derivatives, innovations by Nazi chemists in World War II opened new possibilities through the use of jet fuel. Following the experiments with propeller-based machines begun by the Wright brothers in the 1910s, individual biplanes and pilots squared off in World War I using machine guns and dropping bombs. The era of the Red Baron gave way to stunt flyers known as "barnstormers" who helped to popularize the idea of human flight. By the late 1930s, airlines carried mail and passengers throughout the United States. Flying was expensive and relatively slow; however, Americans who could afford it began taking advantage of its convenience.

World War II helped to commercialize air travel for the United States in a number of ways—for instance, decommissioned military airfields were sold to cities and served as commercial airports, and manufacturers such as Douglas and Boeing remodeled their planes for commercial use, adding pressurized and heated cabins. By the 1950s, millions of Americans used air travel as a viable alternative, and jet engines were also added to the mix by World War II. The first was made in 1931, and the by the 1950s the large-bodied planes took to air, namely the Boeing 747, 767, and 777. Hans von Ohain built a jet engine that flew a plane on August 27, 1939. The engine was powered by gasoline. Englishman Frank Whittle developed his own jet engine, but it wasn't used to fly a plane until May 14, 1941. Because of a gasoline shortage caused by the war, Whittle's engine used kerosene, which remains the base of modern jet fuels. Today, petroleum is refined into a variety of fuels for air travel, including Jet A and Jet A-1.

Using these sources of power, travel and commerce embraced air travel to synch together the globe in a profoundly new fashion by the 1960s. From Federal Express deliveries to diplomatic missions, air travel became the lifeblood of the global economy. This connectivity can be traced to the Boeing 707, the first jet airliner, which began flying between New York and London in 1959 and cut the time needed to cross the Atlantic Ocean to just six hours. In the United States, strict federal regulation kept fares high until 1978, when airlines were deregulated. Although some airlines, focusing on business travelers, kept rates relatively high, the new low-fare carriers (such as Southwest, JetBlue, and AirTran) formed to cater to leisure travelers. Air traffic figures soared from 205 million in 1975 before deregulation, to 297 million in 1980 just after, to 638 million in 2000.[18]

Largely taken for granted by modern humans, these networks for the movement of humans and their goods enabled the borderless era of globalization. Interconnectivity allowed for the expansion of global corporations as well as influential agencies such as the United Nations. The potential implications of this globalization brought serious security issues as well. When airliners fell prey to hijackers or their threats, they became symbols of the developed nations that could be accessed by terrorists—small, nonnation actors that wished to leverage their point of view against global powers. In addition, trade and human movement had always created examples of biological exchange, particularly disease. In the 2008 example of "bird flu," we see just one example of how the complex interconnections moved pathogens to new parts of the world. The expansion of AIDS in the 1980s is also thought to have been exacerbated by the new connections of air travel. Finally, the 2020 spread of COVID-19 is believed to have begun in China and spread quickly through travelers to population centers all over the world.

DEVELOPMENT THROUGH URBANIZATION

In their complexity and resource intensiveness, the modern city dwarfs living environments of the past. "About the only cities," writes McNeill, "that could reach large size before the modern period were imperial centers, whose trajectories waxed and waned with political fortunes, and mercantile centers that depended on overseas trading networks."[19] Even as late as the eighteenth century, only a handful of the world's cities exceeded a population of half a million. As all the factors of the late twentieth century converged, though, urbanization became a clear phenomenon of the Great Acceleration. Larger cities began to emerge before 1940 in Africa, Latin America, and Asia for much the same reasons that they had already seen in Europe and North America, including Cairo expanding to

1.3 million occupants by 1937; Buenos Aires growing to three million by 1950; and Mexico City doubling its size between 1920 and 1920.[20] It was, however, the period after decolonization that began what McNeill terms "the crescendo of urbanization."

In every part of the world after 1950, cities grew faster than rural areas. In 1950 there had been only two cities with populations greater than ten million; by the end of the century such urban monoliths had been termed "megacities" and there were twenty scattered throughout the globe. Such huge concentrations of humans inevitably produced intense stresses on the environment, including air and water pollution, sanitation issues, and disease. In the developed world, between 1950 and 2003 the number of people living in cities more than doubled, from 430 million to 900 million.

Clearly, though, urbanization and particularly the development of the megacity most affected developing nations, where the share of people living in cities more than doubled between 1950 and 2003. By 2003, 42 percent of the population in developing nations lived in cities. In some areas, such a shift reflected a national strategy for modernization, such as China during the Cultural Revolution. In other areas, the mechanization of agriculture left peasant farmers with nowhere else to go but cities. In the Persian Gulf region, for instance, writes McNeill, "the 1973–1974 oil price hikes brought enormous revenues in the region. Showcase cities such as Dubai and Abu Dhabi emerged, characterized by immense wealth and in-migration."[21] Often, existing local governments were overwhelmed by the population influx. For instance, in Bangladesh, Dhaka grew from a small city of four hundred thousand in 1950 to over thirteen million in 2007.

Often the population influx was so rapid that settlements of refugees formed on the swollen cities' outskirts. No suburbs, these settlements teemed with squatters and often held as much as a third of the city's entire population with no infrastructure or sewage to speak of, including nine million outside of Mexico City; three million in Sao Paulo; and over half the population of Mumbai.[22] By the start of the twenty-first century, writes McNeill, "cities in the developing world suffered from the environmental consequences of both extreme poverty and concentrated wealth."[23]

GLOBALIZING THE AUTOMOBILE THROUGH ROADS

In the same effort to modernize that compelled the expansion of cities, nations sought vehicles— the personal transportation device that is most often powered by fossil fuels, specifically petroleum-derived gasoline—for their citizens in the late 1900s. Discussed in earlier chapters, the automobile was at its core a technology reliant on a stable supply of energy—in other words, its use was an indicator of society's energy dependence. The device itself was perfected over the course of the twentieth century, but its foundational innovation was a flexible enough technology that industrialists or governments throughout the globe could create their own version. Even when developed countries were challenged in the 1970s to confront the limits of petroleum supplies, developing nations continued to view the automobile as the first step in remodeling society—varying in scale and scope with each nation—triggering a necessary modernization of infrastructure, including roads, refilling stations, bridges, and new styles of living and shopping patterns. Particularly as less developed nations sought to modernize after colonization, the automobile became an ideal vehicle for change.

Soon after automobiles were mass produced early in the twentieth century, they began to change essential styles of living. Today, the automobile is still causing changes. Easy access by passenger car or by truck helps to determine where people build homes, buy food, seek recreation, and locate businesses. The immediacy and exactness of travel by car and truck make a unique form of transportation. They move near the source or destination of farm or manufactured products, unrestricted by the need for rails, runways, or waterways. Automobility, of course, requires roads, which now cover the industrial countries of the world in a vast network. While automobile culture has evolved throughout the twentieth century, some of the most acute changes occurred during in the 1970s, 1980s, and early 1990s. Concern with safety and pollution led to design changes and the introduction of new technology. Automobile bodies and engines became smaller and lighter to save gasoline, and recently researchers have emphasized alternatives to the gasoline engine.

Whereas the twentieth century began with developed nations prioritizing national road systems, modernizing nations in the twenty-first century continue to see highways as a necessary infrastructure for their own economic expansion. In one of the leading examples, since the 1960s Brazil has approached its rainforest as an impediment to agriculture and economic development. Undertaking massive road building in order to access the forest, about 95 percent of all deforestation occurs within fifty kilometers of highways or roads in the Brazilian Amazon; fragmentation from roads also leads to tree mortality, drought, and onslaught of invasive species.[24] The first major highway in the Amazon basin was cut from the capital Brasilia to Belem at the mouth of the Amazon to encourage settlement into the north of Brazil. In the 1970s, deforestation increased with the beginning of the Trans-Amazonian Highway, built east–west, primarily through the state of Para. Along with this highway came loggers, farm settlers, and land speculation. Forest was often cleared to claim title; cattle grazing allowed newcomers to claim large areas that were lightly inhabited.[25]

By the early 1990s more than fifty million automobiles were produced worldwide annually. Leading manufacturing areas were Japan, the United States, and

Western Europe. Cars had become a mass consumer item after World War II in the United States and Canada, particularly with the advent of massive suburbanization. Between 1950 and 1970, the auto emerged as a consumer item in Western Europe, Japan, and Australia; however, in 1990 Americans still owned more cars than any other society. Even though Japan did not have the space to suburbanize as had the United States, its vehicle ownership increased dramatically after 1960. "In 1990, on average," McNeill writes, "Americans traveled more than twice as far per year in private cars as Europeans, and significantly farther than Australians." And, of course, overall, American manufacturers and consumers preferred larger, heavier vehicles than did other nations.[26]

Road Safety in the Developing World

In many developing nations, a lack of roads was not the only difficulty with assimilating new vehicles and drivers. With little guidance and oversight of roads or driving, traveling by vehicle in many nations has proven quite dangerous. Every year approximately 1.35 million people die as a result of a road traffic crash and between 20 and 50 million more people suffer nonfatal injuries. Around 93 percent of the world's fatalities on the roads occur in low- and middle-income countries, even though these countries have only approximately 60 percent of the world's vehicles. Road traffic injuries are the leading cause of death for children and young adults aged five to twenty-nine years.

The World Health Organization (WHO) has identified traffic accidents as a primary problem in industrializing nations and works with its member states to ensure road safety policy planning, implementation, and evaluation. For example, WHO is currently collaborating with the Bloomberg Initiative for Global Road Safety 2015–2019 to reduce fatalities and injuries from road traffic crashes in targeted low- and middle-income countries and cities. In 2017, WHO released "Save LIVES: A Road Safety Technical Package," which synthesizes evidence-based measures that can significantly reduce road traffic fatalities and injuries. It focuses on **S**peed management, **L**eadership, **I**nfrastructure design and improvement, **V**ehicle safety standards, **E**nforcement of traffic laws, and postcrash **S**urvival. The package prioritizes six strategies and twenty-two interventions addressing the risk factors highlighted above and provides guidance to member states on their implementation to save lives and meet the road safety target of halving the global number of deaths and injuries from road traffic crashes by 2020.[27]

In such a situation, new technologies could emerge suddenly—almost artificially—in societies that otherwise remained less developed. Therefore, in addition to the auto itself, developed nations—"haves"—exported the methods and practices for integrating and operating transportation networks. It also allowed developed nations to guide new adopters of some of the lessons that they had learned.[28]

REGULATING AND MITIGATING VEHICLE EMISSIONS

As scientists began to understand the complexities of air pollution in the late 1960s, it became increasingly apparent that in addition to specific toxic emissions such as lead, the internal combustion engine (ICE) was a primary contributor to air pollution, which in cities is usually referred to as smog. In the United States, for instance, emissions from the nation's nearly two hundred million cars and trucks accounted for about half of all air pollution and more than 80 percent of air pollution in cities. The American Lung Association estimates that America spends more than $60 billion each year on health care as a direct result of air pollution.[29]

When the engines of automobiles and other vehicles burn gasoline, they create pollution. These emissions have a significant impact on the air, particularly in congested urban areas. This is hard to track or trace, though, because the sources are moving. The pollutants included in these emissions are carbon monoxide, hydrocarbons, nitrogen oxides, and particulate matter. Nationwide, mobile sources represent the largest contributor to air toxins, which are pollutants known or suspected to cause cancer or other serious health effects. Greenhouse gases, which are pollutants known or suspected to cause cancer or other serious health effects, are not the only problems, though. Internal combustion engines also emit greenhouse gases, which scientists believe are responsible for trapping heat in the Earth's atmosphere.

Initial efforts at controlling auto emissions date back to 1961 when a single state, California, exceeded anything being considered on the national level and required all cars to be fitted with PCV valves, which helped contain some of the emissions within the vehicle's crankcase. Federal legislation began in 1965 with the Motor Vehicle and Air Pollution Act and was followed in 1970 by the first Clean Air Act. As the new social movement of modern environmentalism took shape after Earth Day 1970, though, constituents forced many lawmakers to consider drastic changes to vehicles.

The organizer of Earth Day, Gaylord Nelson, in fact, went on record in 1970 saying: "The automobile pollution problem must be met head on with the requirement that the internal combustion engine be replaced by January, 1, 1975."[30] Discussed above, the 1973 oil embargo added supply concerns to the calls for the construction of more efficient engines. One of the major proponents of clean air legislation was Senator Edwin Muskie, a Democrat from Maine. He acted as a bridge between the new environmental NGOs springing from middle-class America's Earth Day exuberance and the 1960s conception of using the federal government to regulate and ultimately solve the nation's various ills. Together, a conglomeration of concerns focused public opinion against the internal combustion engine as an inefficient, polluting threat to U.S. health and security. Although Nelson and others argued for banning the engine altogether, the most likely outcome appeared to be placing federal regulations (similar to those used in California) on American cars.

As American policy makers tried to reconcile these new concerns with those of automakers during the 1970s energy crisis, a regulatory framework took shape that is referred to as Corporate Average Fuel Economy (CAFE) standards. As the details were worked out in Congress, Muskie won a major victory when specific pollutants contained in vehicle exhaust, such as CO and HCl, were required to drop 90 percent from 1970 levels by 1975. The intention, of course, was to force manufacturers to create the technologies that could meet the new standards. Individual states led the way. In 1975, a California act required that vehicle exhaust systems be modified prior to the muffler to include a device, the catalytic converter. Costing approximately $300, early converters ran the exhaust through a canister of pellets or honeycomb made of either stainless steel or ceramic. The converters offered a profound, cost-effective way of refashioning the existing fleet of vehicles to accommodate new expectations on auto emissions.

In addition, the scientific scrutiny of auto emissions proceeded on one additional, much more specific front. Air testing on emissions and the smog that they created also revealed a now undeniable reality of auto use: lead poisoning. The willingness to tolerate lead additives in gasoline had persisted from the 1920s. Under the new expectations of the 1970s, though, lead emissions presented auto manufacturers with a dramatic change in the public's expectations. By this point, the amount of lead added to a gallon of gasoline hovered in the vicinity of 2.4 grams. In January 1971, EPA's first administrator, William D. Ruckelshaus, declared that "an extensive body of information exists which indicates that the addition of alkyl lead to gasoline . . . results in lead particles that pose a threat to public health." The resulting EPA study released on November 28, 1973, confirmed that lead from automobile exhaust posed a direct threat to public health. As a result, the EPA issued regulations calling for a gradual reduction in the lead content of the nation's total gasoline supply, which includes all grades of gasoline.[31]

Given the degree of regulation and the immense, new expectations placed on vehicles, American auto manufacturers came out of the 1970s feeling under siege. Each leader in the industry forecasted expensive shifts that would raise vehicle prices and put American laborers out of work. In fact, some openly speculated about whether automobiles could hope to still be manufactured in the United States in the twenty-first century. They would apply their considerable creativity to extending the American tradition of car-making into the next century; however, American manufacturers obviously directed this creative design toward circumventing new regulations. In truth, though, air pollution was simply an immediate outcome of burning fossil fuels. The implications and impacts of these emissions would prove even more troublesome to the future of the ICE.

GREEN REVOLUTION AND GLOBAL FOOD SCIENCE

Human movement was not the only characteristic of everyday life affected by the Great Acceleration. Food preparation and availability significantly changed with the widespread availability of inexpensive energy. Simplifying and increasing agricultural productivity may have been petroleum's clearest path toward emerging as more than a fuel. Systems of trucking for dispersal and mechanized tools such as tractors and combines, each powered by diesel fuel, transformed humans' food network and, in most cases, significantly reduced food prices. Using petroleum as a fuel, however, was only a portion of the revolution that crude brought to food patterns in the twentieth century. Chemists also played an important role in using petroleum to help expand, simplify, and, at times, enhance the food that appeared in our shopping carts.

Such petrochemical breakthroughs enabled agricultural nations to expand and simplify agriculture, creating a role for modern agricultural corporations throughout the world. In addition, however, these methods and chemicals stimulated one of the first great episodes of international cooperation, known as the "Green Revolution." Reaching across "the gap," agricultural scientists from the developed world during the 1960s shared techniques and hybrid seeds with farmers in less developed nations. The Green Revolution brought agricultural technology to many civilizations in which farming was failing for one reason or another, and, therefore, it brought food to starving people in Africa and South America. This revolution in global agriculture began with wheat crops in Mexico and then India before expanding into maize in Africa.[32]

In the process of this assistance, however, many of the techniques and products incorporated these farmers in a model of agriculture much more reliant on petroleum than their previous methods. This was most evident in the use of agricultural chemicals such as industrial fertilizers, the use of which increased seven-fold between 1975 and 2000.[33] Geomorphologist David Montgomery writes that "for the period 1961–2000, there is an almost perfect correlation between global fertilizer use and global grain production."[34] In many of the nations of Asia and Central and South America, the Green Revolution has been a terrific success in food production—in Asia, for instance, more than three-quarters of the rice grown is from introduced crops. In other areas, such as Africa, it has been less successful.

PETRODICTATORS AND SOCIALISTS
LEVERAGE GROWING SCARCITY

In an era of petroleum hoarding, nations possessing crude—the "haves"—obviously experienced a significant increase in their global stature. The term *petrodictator* has been attached to a variety of leaders in locations ranging from Azerbaijan to Venezuela. In such cases, petroleum reserves have not been used by the domestic population and instead leaders have leveraged the profit that they can bring on the global market. Some leaders have used oil profits to raise the international stature of their nations; others have used such wealth for personal enrichment while the nation languishes economically. Although part of the landscape that holds "new oil wars," these petrodictators are not only using their oil for its rentier value. Defining the form, Saddam Hussein, the Iraqi leader from 1979 to 2003, pressed the advantage of petroleum wealth more than any other (Iraq had nationalized its oil industry in 1972). In the end, most observers would claim that he overplayed his petroleum advantage; petrodictators who followed have learned from his example how to preside over the commodity in this new era of resource wars.

In his boldest move, of course, Hussein sought to function as the enforcer of the OPEC interest to limit production and, thereby, manage the global price. Making its own determination to appease the United States, Kuwait leaders consistently overshot its production caps during the 1980s. Although this willingness enhanced the nation's relations with Western powers, OPEC leaders grew increasingly frustrated with its rogue production. In Iraq, Hussein rose to dictatorial power in 1979 and began building the region's largest military. First used in 1979 to invade Iran, in 1990 Hussein, with an additional eye toward Kuwait's access to the Persian Gulf, decided to be OPEC's enforcer and the Iraqi army invaded Kuwait. Increasing his oil reserves by 20 percent overnight, the world looked on and imagined the consequences if Hussein's campaign continued into the lightly armed Saudi Arabia and United Arab Emirates—which would then provide him with control of approximately half of the world's proven petroleum reserves. With its hand forced by a continued need for petroleum, the United States and allies "drew a line in the sand."

The ensuing months were marked by efforts to use the United Nations to arrive at a nonmilitary diffusion to the situation. This was abandoned, ultimately, on January 17, 1991, when a massive multinational force authorized by the UN and led by the United States descended on the Persian Gulf with the goal to return Hussein and his army to Iraq. Most of the fighting lasted just hours as Hussein's army suffered grave defeat at the hands of the world's most advanced military technology. In fact, the war itself symbolized the gap between developed and less developed nations as Hussein's army—dominant within the Middle East and African region—appeared primitive and hopelessly overmatched. This, however, did not mean that his forces could not exert great damage on the real source of the conflict: before retreating, Hussein's army set afire approximately eight hundred Kuwaiti oil wells, creating a modern environmental disaster.

Allowed to retreat to Baghdad, Hussein remained in power until another American president—another George Bush—seized the moment in 2003 to commit a largely American force to dislodging the dictator. The logic of the war in 2003 was tied to the attacks on American soil on September 11, 2001. Action against leaders such as Hussein, President George W. Bush argued, fell into a new strategy of "pre-emptive" warfare that was designed to head off future attacks or conflicts. Critics immediately claimed it was a resource war designed to open Iraq's petroleum reserves to unfettered use and development by the United States. For the purpose of our consideration, hindsight demonstrates that at the very least access to crude was one of the fringe benefits to unseating Hussein. The instability that ensued after Hussein's fall, capture, and death thwarted hopes for immediate development of Iraqi oil; however, by 2010, Iraq's new petroleum order was clear.

Learning from Hussein's model, the next most obvious petrodictator was Venezuela's Hugo Chávez. A great admirer of Cuba's Fidel Castro, Chávez swept to political leadership in 1998 and, ever since, has sought to use his nation's enormous oil reserves to leverage international standing for himself and Venezuela. Internally, Chávez promised "revolutionary" social policies and constantly abused the "predatory oligarchs" of the establishment as corrupt servants of international capital. Internationally, Chávez employed what he refers to as "oil diplomacy." Venezuela has "a strong oil card to play on the geopolitical stage," he explained. "It is a card that we are going to play with toughness against the toughest country in the world, the United States."[35] In OPEC, Chávez has fought to keep prices high and has even publicly questioned whether barrel prices should still be measured on the basis of the American dollar. Whether speaking at the United Nations to demonize the United States or threatening to only sell their oil directly to underprivileged populations in the United States, Chávez's international standing—whatever it might actually be—is based on his nation's vast supply of crude.

Russia has followed a different political model in recent years; however, petroleum has emerged as a major structuring agent for its base of national power following the fall of communism. Oil production is no longer financed by the state budget but by selling the output to other nations. In at least one region—western Siberia—just as the communist government fell and Russia emerged as an independent region, the former Soviet Ministry of Oil petitioned Moscow to form a joint stock company known as Lukoil.[36] Other petroleum resources were divided among workers and private companies in very complex and unclear arrangements during the early days of Russia's independence. Historian John D. Grace writes: "By the beginning of 1995, of the roughly three dozen original

Soviet-era producers in Russia, over 20 were still wholly in state hands and 13 were listed as private companies. . . . The most important of these were Lukoil, Yukos, Surgutneftegaz, Slavneft, Sidanco, Kominift, Eastern Oil and Onako." In the Volga-Ural basin, Grace added two companies that remained under the control of local governments: Tatarstan and Bashkortostan.[37] As a few Russians took control of the nation's banking system, these "oligarchs" soon became major players in the new oil companies—particularly in Lukoil.

By the early twenty-first century, Lukoil used Western oil and gas corporations as its model. It took over smaller companies and diversified into international operations beyond exploration and production, including refining, marketing, and the petrochemical industry. In 2002, Lukoil became the first Russian oil company to list its shares on a Western exchange (in London).[38] Lukoil became an active player in Colombia and Iraq and also took over many of the major pipeline projects near the Caspian Sea. New trading arrangements were formed with Asian nations, particularly Japan and China, and poised Lukoil to take advantage of some of the world's fastest growing oil markets in the twenty-first century. Whether the companies are truly independent or not, thanks to their rapid success the new Russia stands as a leader in production and distribution of oil today.

Petrodictators have often managed to maintain control of their nations for fairly lengthy regimes. There is growing evidence, however, that a lopsided emphasis on petroleum development by dictatorial powers does not end well. Regardless, petroleum supplies have emerged as the single most significant equalizer for nations on the less developed side of the gap discussed above.

Other nations have used a state-owned or socialist model to emphasize oil development to support infrastructure development. In developing the North Sea supply of oil, for instance, a group of European nations has carried out a joint initiative. The United Kingdom, Denmark, Norway, Germany, and the Netherlands formed joint tax and licensing regimes to develop the difficult North Sea offshore supply after 1968. In Norway, for instance, the state-owned Statoli corporation has helped the nation become the world's third largest oil exporter and eighth largest producer. Choosing not to join OPEC, Norway instead established the Petroleum Fund of Norway in 1990 to collect profits from sales and licensing fees. One of the largest public funds in the world, this fund is largely held to ensure the nation's economic stability when oil supplies diminish. Particularly because Norway's population stands at less than five million, critics in recent years have questioned whether it is necessary to create such a large savings fund. Recently, many critics have called for the fund to be used more for internal improvements and national needs.

WARRING FOR PETROLEUM ACCESS

From the perspective of developed nations on the other side of the gap, the use of petroleum supplies as a political weapon demanded an increasingly active culture of engagement. At his inauguration as U.S. president in 1989, George H. W. Bush seemed to speak directly to the Middle East and to petrodictators when he said: "They got a president of the United States that came out of the oil and gas industry, that knows it and knows it well." Bush's worldview teamed with his business experience to make him one of the first Western leaders who clearly—and openly—believed in the strategic importance of a U.S. influence in the OPEC-dominated Middle East, which was now responsible for producing two-thirds of the world's oil.

At this historic juncture, OPEC was wrestling with the idea of fixing petroleum prices for the good of all its members, but many individual nations were unwilling to limit production due to their own economic limitations. When Hussein invaded Kuwait in 1990, American president and oilman Bush orchestrated the joint action by UN forces to stop his progress and, ultimately, force Iraqi troops out of their neighboring nation. As Iraqi forces fled Kuwait, they lit on fire many of the nation's oil wells. This act of terrorism created an environmental hazard and debilitated Kuwait's immediate ability to produce oil. Most damaging, though, was Hussein's miscalculation that presented the United States with a military presence in the world's oil region. Bush accomplished his goal of creating a mutually dependent relationship between Persian Gulf nations and the United States. However, this did not necessarily mean that price stability would last. The late 1990s brought more problems related to underproduction. The production imbalance fed the tripling of gasoline prices in 1999–2000. As Kaldor and colleagues trace the roots of the twenty-first-century war in Iraq, the United States, they argue, sought to fight an "old war" about oil by using the commodity to facilitate the decision to go to war. The conflict, it was argued, would be paid for and largely absorbed by Iraq's "oil revenues."[39]

In a twist of historical fate, the American presidential election of 2000 brought George W. Bush, son of the previous president, into office. Although Iraq's leader Hussein and Middle Eastern oil supplies were priorities of the younger Bush, oil prices remained somewhat low. Energy security, though, emerged in the public sphere with the 9/11 attacks of 2001. Although unrelated to Hussein, these attacks became a leveraging point with which President Bush could make unseating the Iraqi leader a mission of the American military in two fashions: first, his unreliability and possible dangerousness was compounded by his control of such significant oil reserves; second, if an invasion was carried out, it was argued, revenue from petroleum sales would help quickly stabilize the new Iraq and decrease the financial resources necessary from the United States or any other occupying nation. Each of these components for war derived from the importance of petroleum.

Had the 2003 American-led invasion been about oil supplies? Wound tightly into the strategic need for energy, developed nations required a steady supply of crude. And, clearly, the need to preserve energy security abroad had steadily increased during the twentieth century. Until the end of World War II, domestic supplies allowed Americans to watch European powers colonize and develop Middle Eastern supplies; the end of the war, though, brought the same realization to American foreign policy makers. The world that emerged in the twenty-first century clearly factored geopolitics into nearly every diplomatic interaction. Protected from this reality by over a half-century of disinformation, American consumers clearly became the last remaining disconnect in comprehending the implications of our dependency. History may show that the invasion, occupation, and support of Iraq had crude at its core. Clearly, however, this war was a leading symbol of a new world petroleum order organized by the haves and have-nots and various efforts to compensate for each nation's particular standing in the petroleum organization.

To make the new petroleum order even more clear, in 2004, Nancy Birdsall and Arvind Subramanian published "Saving Iraq from Its Oil" in the influential journal *Foreign Affairs*. The article sought to advise the United States and occupying nations how they might best make Iraq's petroleum a beneficial resource for developing a new nation. This argument was based on a simple yet remarkable main idea. They write of petroleum's "resource curse" in this fashion:

> Oil riches are far from the blessing they are often assumed to be. In fact, countries often end up poor precisely because they are oil rich. Oil and mineral wealth can be bad for growth and bad for democracy, since they tend to impede the development of institutions and values critical to open, market-based economies and political freedom.[40]

The bitter irony of the Iraq War, of course, is that despite its origins, it became a "new oil war" in which nonstate actors took an active role in fomenting dissent and complicating American occupation of the country. In the process, these activities by terrorists neutralized the ability of profits from crude to assist the settlement of a new Iraq.

DUBAI AND HUMAN MIGRATIONS FOR CRUDE

Symbols can be very important to entire societies, particularly when it comes to their attempts to create an image as an independent, modernizing state. For instance, the events of 9/11 demonstrated that for some in less developed nations, made to feel powerless by a lack of opportunity and access to developing their nation's resources, two identical office towers, known as the World Trade Centers in New York City, marked the point at which—in some small or potentially significant way—the gap could be breached. Although the loss of thousands of American lives that day has resulted in expanded military activity by the United States as well as an intensified culture of domestic security, the world's skyline tells us that the gap has very likely not closed; however, the centers of power may have begun to shift. In this case, while the vast majority of the population of the United Arab Emirates (UAE) lives in intense poverty and lacks basic aspects of modern life, developing Dubai as a global center takes precedence.

Erupting into the sky with much more symbolism than utility, the Burj Khalifa opened for business in Dubai in early 2010. A symbol of an emerging world order, Burj Khalifa is a rocket-shaped edifice that soars 828 meters, or 2,717 feet. It is the world's tallest structure, with views that can reach a hundred kilometers (approximately sixty miles). At a cost estimated at $1.5 billion, the Burj took five years to build, is over 160 floors high, and has comfortably surpassed the previous record holder in Taipei. If the 1970s oil crisis marks the point where developed nations were forced to acknowledge their need for oil from less developed nations, Burj Khalifa marks the permanent institutionalization of this priority.

Dubai, the city that the Burj towers over, emerged in the first decade of the twenty-first century as a global phenomenon. It is one of the seven emirates of the UAE, located south of the Persian Gulf on the Arabian Peninsula. The Dubai municipality is sometimes called "Dubai state" to distinguish it from the emirate. Although the city grew with the petroleum industry at the close of the twentieth century, at the start of the twenty-first Dubai positioned itself as the economic center for an emerging global economy. Focused on organizing the financial development of projects in the Middle East and Southeast Asia, the city also functions as an oasis for the diverse workforce that began to operate within the Persian Gulf.

The increasing importance of energy was noticeable not only in political leadership at the dawn of the twenty-first century. Harvesting crude, wherever it occurred, also created patterns of worker migration from neighboring regions. In the Persian Gulf region, population shifts connected to petroleum have become a defining characteristic for the entire region. One of the most significant labor influxes to the Persian Gulf since 1990 has been from Kerala, India. In 1998, for instance, nearly 1.4 million Keralans emigrated from India, approximately 95 percent of whom were destined for Arab countries of the Middle East. Nearly 40 percent of this total immigrated to Saudi Arabia and 30 percent to the UAE.

In the Middle East, generations of workers have moved with the oil industry. Facing low-paying jobs or unemployment at home, for decades many have worked abroad as laborers, taxi drivers, or food preparers, wiring money to their families or returning with it on occasional visits. Recent years have seen an increase in the level of professional workers, particularly coming from nations such as Egypt. Whereas in the past, engineers and other professionals would take their training to find employment to Europe or the United States, they are finding increasing opportunities closer to home. Egypt has an estimated five million workers abroad, including 1.5 million based in the gulf region. Remittances from those in the United States, Europe, and the Persian Gulf are a key source of foreign currency for Egypt. Egyptians sent home $8.56 billion in remittances in the 2007–2008 fiscal year, up from $6.32 billion a year earlier.

Much of this work is focused on a new city that has taken shape in the region: both in the work of international economic trading that takes place in it and in constructing the physical monument in which much of it will take place. Blending World Trade Center with Las Vegas decadence, Dubai now focuses around the Burj, with its mix of nightclubs, mosques, luxury suites, and boardrooms. In the Burj, one finds the extravagant splendor of the world's first Armani hotel, the world's highest swimming pool (on the 76th floor), the highest mosque (on the 158th floor), and fifty-four elevators that can hit speeds of sixty-five kilometers (40 miles) per hour. For the more than twelve thousand people who occupy its six million square feet, the Burj is an oasis from the desert that surrounds it as well as from the overwhelming poverty of the majority of the UAE public.[41]

CONCLUSION: REINFORCING THE ENERGY GAP

Business began on September 23, 2008, but it was anything but "as usual." In such a moment, setting up an office functions on a literal office: buying desks, copiers, and paper clips; in some cases, though, it also functions on a symbolic level. In the case of the office of Shell Oil opening on this day in Baghdad, Iraq, it functioned on a series of symbolic levels. For instance, although its intention was to conduct business, its location needed to be kept secret. If its location was revealed, the office would likely be attacked by Al-Qaeda or other enemies in the unstable, American-occupied nation. Therefore, it was a symbol of the precarious occupation undertaken by the United States five years earlier. However, the opening of the Shell office also marked the reintroduction of foreign oil companies following a thirty-two-year absence. Shell's office suggested a reality of the invasion that the American administration of George W. Bush had long denied: petroleum had been a serious consideration for its 2003 invasion of Iraq.

Petroleum as an issue of national security would have never occurred to the American public in 1950, even though World War II had made political leaders acutely aware of its strategic importance; however, today, the concept is so common that most Americans find it impossible to recall a day when they controlled the bulk of the world's petroleum supply. Indeed, a new world order has evolved that is organized by the power emanating from crude—those who must have it but do not possess sufficient reserves themselves and others who possess it but do not need it for themselves. In this twenty-first-century era of haves and have-nots, political scientist Michael Klare writes that the United States and other developed nations now exist in an era of "resource wars," which he describes in this fashion:

> For the American military establishment, this concern has particular resonance: while the military can do little to promote trade or enhance financial stability, it *can* play a key role in protecting resource supplies. Resources are tangible assets that can be exposed to risk by political turmoil and conflict abroad—and so, it is argued, they require physical protection. While diplomacy and economic sanctions can be effective in promoting other economic goals, only military power can ensure the continued flow of oil and other critical materials from (or through) distant areas in times of war and crisis. As their unique contribution to the nation's economic security, therefore, the armed forces have systematically bolstered their capacity to protect the international flow of essential materials.[42]

Economists have also parceled the concept of resource wars into categories of new and old warfare. By doing so, they follow the logic of this chapter to argue that oil and war have been linked since the start of the twentieth century as oil "was considered a key strategic commodity and security." Economists Mary Kaldor, Terry Lynn Karl, and Yahia Said then go on to explain that in new oil wars, the government connection has been eroded. "New wars," they write,

> are associated with weak and sometimes ungovernable states where non-oil tax revenue is falling, political legitimacy is declining and the monopoly of organized violence is being eroded. In such wars, the massive rents from petroleum are used in myriad ways to finance violence and to foster a predatory political economy.[43]

As a "rentier war," conflict over oil is based only on crude's remarkable value. Interested parties express little or no interest in long-term development of the region or resource. In addition, they often care little about the global nature of the commodity—except that it will bring them revenue. Often, they work with global oil corporations in an unfettered and unregulated arrangement that is seen as a major threat to the stability of crude as a commodity.

With the erosion or entire retreat of colonial authority, many nations suddenly faced power vacuums that were seized by a variety of leaders. The late twentieth century saw many examples of leaders of the developed world attempting to find diplomatic or military methods for managing relations with such individuals. These efforts grew more intense if the nation was of strategic importance because of its location or the resources that it possessed. In this fashion the postcolonial era saw nations categorized under a petroleum measuring stick: have or have-not. Of course, the suddenly independent nations of the Middle East, such as Saudi Arabia, UAE, and Iraq, fell into the "have" category.

Possessing oil, however, did not result in an automatic economic shift within a nation. As journalist Peter Maass writes in *Crude World:* "One of the ironies of oil-rich countries is that most are not rich, that their oil brings trouble rather than prosperity."[44] In Nigeria, government ministers clash with military generals and civilians are entirely ignored in the effort to ease access to the nation's oil reserves. Ecuador's lack of concern over the behavior of oil developers contaminated a tributary of the Amazon River on which all life in the region depends. And, in nations ranging from Russia to Venezuela and to Guinea, government officials have used oil to consolidate political power and undergird their presence on the world stage.

By early 2010, the primary purpose behind the opening of the Shell office in 2008 emerged, as the Anglo-Dutch company and the Malaysian state-run oil company Petronas received a twenty-year deal to develop Iraq's largest oil field, Majnoon. In addition, on January 14, 2010, the headline in the *New York Times* read: "U.S. Companies Race to Take Advantage of Iraqi Oil Bonanza." The familiar large American oil field supply companies (including Halliburton, Baker Hughes, Weatherford International, and others) sought, in the words of the *Times*, "to revive the country's stagnant petroleum industry, as Iraq seeks to establish itself as a rival to Saudi Arabia as the world's top oil producer."[45] Were these elements of Big Oil simply stepping into a vacuum in order to help all concerned by stabilizing Iraq's immense oil wealth? Or had this always been the intention of the U.S. decision makers? Did they make war for oil? An affirmative response coheres with the general trend of the developed world's shift into a more complex culture of petroleum, specifically, and energy supplies in general.

NOTES

1. Jeffrey Liker and Michael Hoseus, *Toyota Culture* (New York: McGraw Hill, 2008), p. 34.

2. See, for instance, Kenneth Pomeranz, *The Great Divergence* (Princeton, NJ: Princeton University Press, 2001).

3. Simon Pirani, *Burning Up* (New York: Pluto Books, 2018), p. 130.

4. Pirani, p. 107.

5. Daniel Yergin, *The Prize* (New York: Free Press, 2008), p. 612.

6. Although Mexico wrestled control of its oil industry from foreigners in 1938, it quickly receded from the lucrative international market due to insufficient capital for investment. Other nations also attempted to set up their own arrangements for oil development, including Venezuela, which in 1943 signed the first "fifty-fifty principle" agreement that provided oil producers with a lump sum royalty plus a fifty-fifty split of profits, and Iran, which passed a law demanding the termination of previous agreements with Anglo-Iran (referred to as Anglo-Persian prior to 1935 and British Petroleum after 1954) and then nationalized its oil operations in 1954 when such an agreement failed to occur. (A new British-Iranian agreement was signed the following year. The newly restored Shah of Iran became a pillar of American Middle East policy until the Iranian Revolution in 1979.)

7. See Fiona Venn, *The Oil Crisis* (London: Routledge, 2017).

8. Venn.

9. *New York Times* February 5, 1974.

10. *New York Times*, January 21, 1974. See Karen R. Merrill's collection, *The Oil Crisis of 1973–4* (New York: St. Martin's 2007).

11. Andrew Scott Cooper, *Oil Kings* (New York: Simon & Schuster, 2011), p. 163.

12. Amory Lovins, *Soft Energy Paths* (New York: Friends of the Eath International, 1977), pp. 121–122.

13. Daniel Horowitz, *Jimmy Carter and the Energy Crisis of the 1970s* (Boston: Bedford Books, 2005), pp. 20–25.

14. Horowitz, pp. 43–46.

15. For more information about SPR, see Bruce Beaubouef, *The Strategic Petroleum Reserve* (College Station: Texas A&M Press, 2007).

16. Shane Hamilton, *Trucking Country* (Princeton, NJ: Princeton University Press, 2008), pp. 100–101.

17. Hamilton, p. 115.

18. For more information, see: http://www.centennialofflight.gov/essay/Social/impact/SH3.htm. Accessed on January 6, 2022.

19. John R. McNeill, *Something New Under the Sun: An Environmental History of the Twentieth-Century World* (New York: Norton, 2001), pp. 106–107.

20. McNeill, p. 112.

21. McNeill, p. 124.

22. McNeill, p. 125.

23. McNeill, p. 126

24. Global Forest Atlas, Yale University: https://globalforestatlas.yale.edu/amazon/land-use/roads-amazon-basin. Accessed on January 6, 2022.

25. Global Forest Atlas.

26. McNeill, pp. 121–122.

27. WHO road traffic injuries: https://www.who.int/news-room/fact-sheets/detail/road-traffic-injuries. Accessed on January 6, 2022.

28. https://www.theglobaleconomy.com/rankings/roads_quality/

29. Jack Doyle, *Taken For a Ride* (New York: Four Walls Eight Windows, 2000) p. 134.

30. Doyle, p. 64.

31. James Motavalli, *High Voltage* (New York: Rodale Press, 2011), p. 40.

32. James McCann, *Maize and Grace* (Cambridge: Harvard University Press, 2007).

33. Christian Smedshaug, *Feeding the World in the 21st Century* (New York: Anthem Press, 2010), p. 221.

34. David Montgomery, *Dirt* (Berkley: University of California Press, 2012), p. 197.

35. Justin Blum, "Chavez Pushes Petro-Diplomacy," *Washington Post*, November 22, 2005. Retrieved November 29, 2005.

36. John D. Grace, *Russian Oil Supply* (London: Oxford University Press, 2005), p. 105.

37. Grace, p. 107.

38. Grace, pp. 110–112.

39. Kaldor et al., *Oil Wars* (London: Pluto Press, 2007), pp. 6-8.

40. Nancy Birdsall and Arvind Subramanian, "Saving Iraq from Its Oil," *Foreign Affairs* 83, no. 4 (Jul–Aug, 2004), p. 77.

41. See, for instance, Jim Krane, *City of Gold: Dubai and the Dream of Capitalism* (New York: Picador Press, 2010).

42. Michael Klare, *Blood and Oil* (New York: Metropolitan Books, 2004), p. 9.

43. Kaldor et al., pp. 2–4.

44. Peter Maass, *Crude World* (New York: Vintage Books, 2009), p.3.

45. Anthony Garavente, "U.S. Companies Race to Take Advantage of Iraqi Oil Bonanza," *New York Times*, January 14, 2010.

IV

INTEGRATING SUSTAINABILITY (2000–2022)

The layers of energy history run deep in a place such as my home in central Pennsylvania in the United States. Our ridges held some of the nineteenth century's greatest stores of coal, which was used to power many industries, but most particularly railroads. From the nexus of Altoona, coal- and diesel-powered train engines ran on tracks that bound together the nation and stimulated economic development in the twentieth century. Today, these ridges house strings of wind turbines that often have been built on former coal-mine lands, such as this one near my home in Hollidaysburg.

In *The Sixth Extinction*, Elizabeth Kolbert writes of the moment in which we now live and the emergence of the Anthropocene concept in this fashion:

Right now, in the amazing moment that to us counts as the present, we are deciding, without quite meaning to, which evolutionary pathways will remain open and which will forever be closed. No other creature has ever managed this, and it will, unfortunately, be our most enduring legacy.[1]

Humans have lived more aggressively than any species in Earth's history; however, correspondingly, through science and technology we have raised our level of awareness and understanding to begin to comprehend and appreciate our impact. With the context that the Anthropocene provides, we can see that the next phase in humans' energy development is one that causes as few additional problems as possible. Someday soon these more benign sources of power might no longer even be known as "alternatives."

NOTES

1. Elizabeth Kolbert, *The Sixth Extinction* (New York: Picador Books, 2015), pp. 268–269.

TRANSITIONING BY THE NUMBERS
Considering Sustainable Energy

In our current transition, each source of energy is priced anew in a complex marketplace that factors in costs and benefits resulting from the use of each prime mover. Most important, carbon accounting no longer allows fossil fuels to singularly appear as the simple, cost-effective path forward.

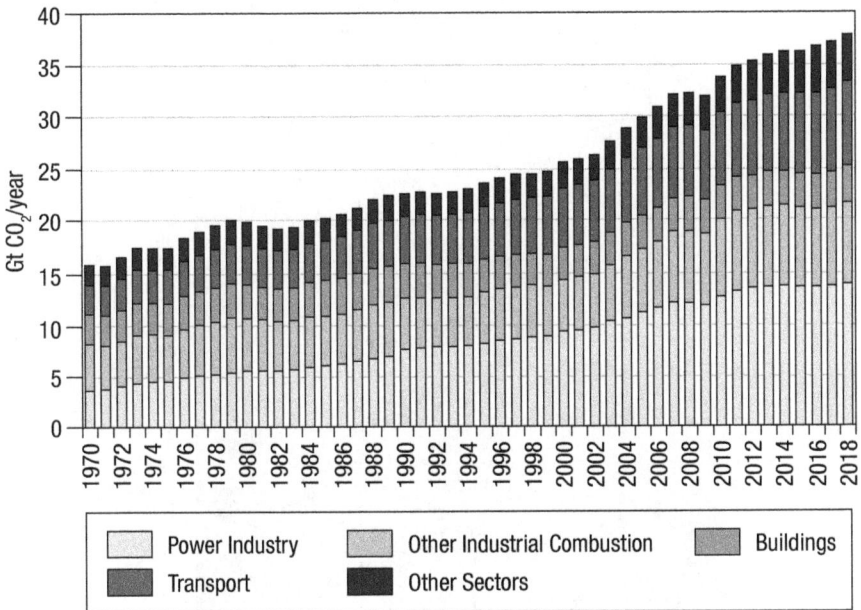

Total global annual emissions of fossil CO_2 in Gt CO_2/yr by sector. Fossil O_2 emissions include sources from fossil fuel use, industrial processes, and product use. SOURCE: RESEARCHGATE: HTTPS://WWW.RESEARCHGATE.NET/FIGURE/TOTAL-GLOBAL-ANNUAL-EMISSIONS -OF-FOSSIL-CO$_2$-IN-GT-CO$_2$-YR-BY-SECTOR-5_FIG1_345685694. ACCESSED JANUARY 6, 2022.

EON	ERA	PERIOD	EPOCH		Ma
Phanerozoic	Cenozoic	Quaternary	Holocene		
					0.011
			Pleistocene	Late	0.8
				Early	2.4
		Tertiary (Neogene)	Pliocene	Late	3.6
				Early	5.3
			Miocene	Late	11.2
				Middle	16.4
				Early	23.0
		Tertiary (Paleogene)	Oligocene	Late	28.5
				Early	34.0
			Eocene	Late	41.3
				Middle	49.0
				Early	55.8
			Paleocene	Late	61.0
				Early	65.5
	Mesozoic	Cretaceous	Late		99.6
			Early		145
		Jurassic	Late		161
			Middle		176
			Early		200
		Triassic	Late		228
			Middle		245
			Early		251
	Paleozoic	Permian	Late		260
			Middle		271
			Early		299
		Pennsylvanian	Late		306
			Middle		311
			Early		318
		Mississippian	Late		326
			Middle		345
			Early		359
		Devonian	Late		385
			Middle		397
			Early		416
		Silurian	Late		419
			Early		423
		Ordovician	Late		428
			Middle		444
			Early		488
		Cambrian	Late		501
			Middle		513
			Early		542
Precambrian	Proterozoic	Late	Neoproterozoic (Z)		1000
		Middle	Mesoproterozoic (Y)		1600
		Early	Paleoproterozoic (X)		2500
	Archean	Late			3200
		Early			4000
	Haydean				

This map of geologic epochs lays out Earth history prior to our entry into the current Anthropocene. SOURCE: NPR, "CLIMATE CHANGE AND THE ASTROBIOLOGY OF THE ANTHROPOCENE." HTTPS://WWW.NPR.ORG/SECTIONS/13.7/2016/10/01/495437158/CLIMATE-CHANGE-AND-THE-ASTROBIOLOGY-OF-THE-ANTHROPOCENE. ACCESSED JANUARY 6, 2022.

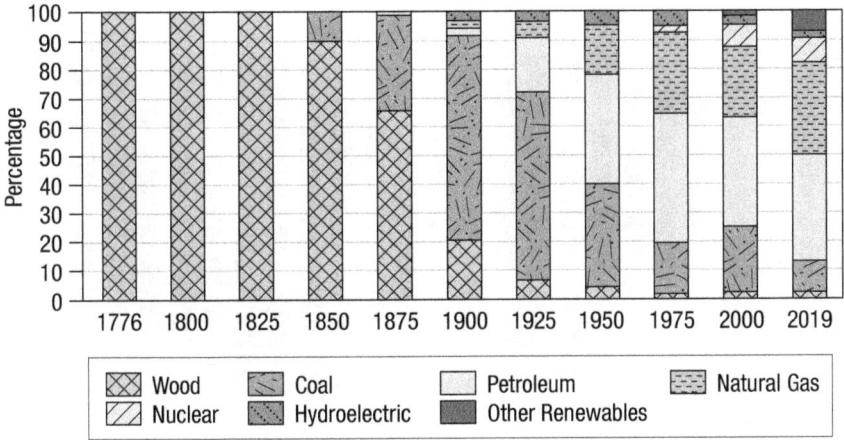

Diversifying Energy Use in the United States, 1776–2019. SOURCE: BASED ON IEA DATA FROM IEA (2019) MONTHLY ENERGY REVIEW, IEA (2020), WWW.IEA.ORG/STATISTICS. ACCESSED JANUARY 6, 2022. ALL RIGHTS RESERVED; AS MODIFIED BY BRIAN BLACK.

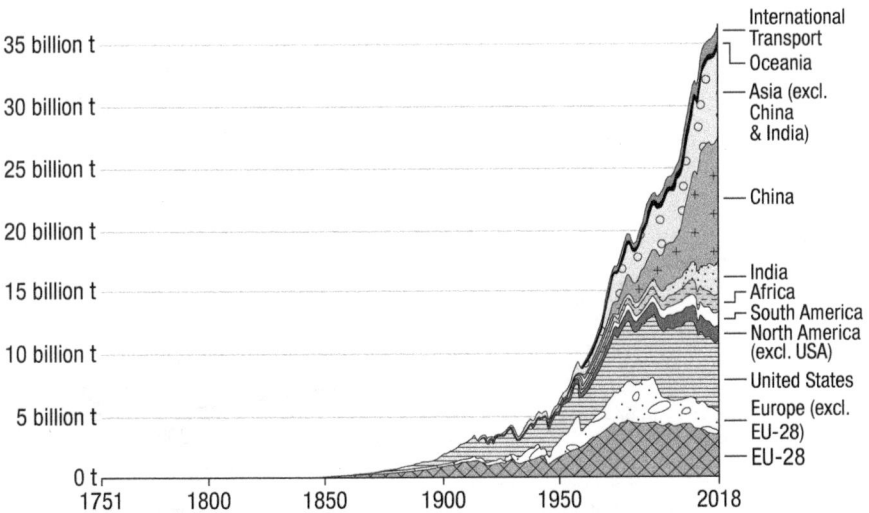

CO_2 Emissions through 2018 by World Region. SOURCE: CARBON DIOXIDE INFORMATION ANALYSIS CENTER (CDIAC): GLOBAL CARBON PROJECT (GCP). HTTPS://OURWORLDINDATA.ORG/ GRAPHER/ANNUAL-CO-EMISSIONS-BY-REGION. ACCESSED JANUARY 6, 2022.

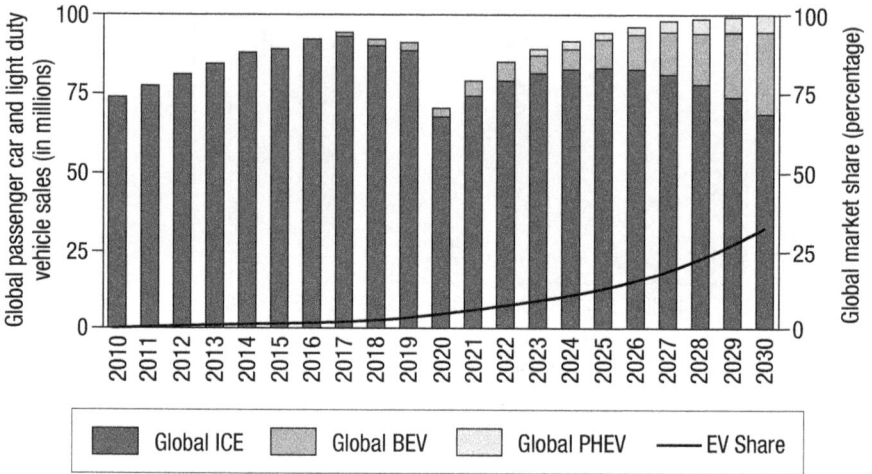

Outlook for Global Passenger Cars to 2030. SOURCE: DELOITTE, "ELECTRIC VEHICLES: SETTING A COURSE FOR 2030": HTTPS://WWW2.DELOITTE.COM/UK/EN/INSIGHTS/FOCUS/FUTURE -OF-MOBILITY/ELECTRIC-VEHICLE-TRENDS-2030.HTML. ACCESSED JANUARY 6, 2022.

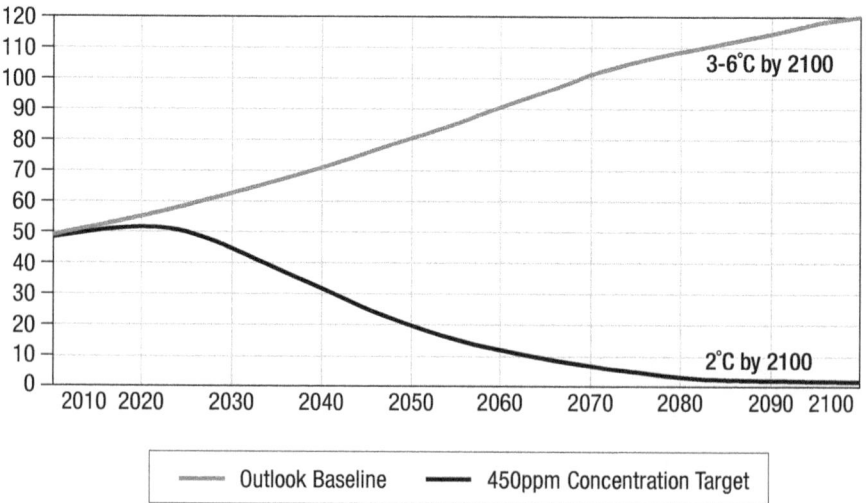

Global Emissions with Mitigation Targets, 2010–2100. SOURCE: OECD, CLIMATE CHANGE MITIGATION: HTTPS://WWW.OECD.ORG/STATISTICS/CLIMATE-CHANGE-MITIGATION-WE-MUST -DO-MORE.HTM. ACCESSED JANUARY 6, 2022

8

Energy Transitions and the Culture of Sustainability

GROUND: OPEN COMPETITION FOR ARCTIC ENERGY RESOURCES

Is it a new, very cold war? Russia, China, and the United States (along with a few other nations) are openly vying for influence and control in the Arctic. None of the interest derives from any desire for long-term settlement; instead, nations are eyeing $35 trillion worth of untapped oil and natural gas that experts believe could be accessible there. Although the warming climate means difficulty for much of the world, these nations see it as an opportunity in this part of the world.

With more than half of all Arctic coastline along its northern shores, Russia has long sought economic and military dominance in this part of the world. In recent years, warming global temperatures opened up new sea lanes and economic opportunities at the top of the world, and China initiated a "Polar Silk Road" to try to take advantage of the largely unowned reserves. Although the region is thought to hold gold, silver, diamond, copper, titanium, graphite, uranium, and other valuable rare earth elements, it is the energy resources that attract most attention: experts forecast that the Arctic could hold between one-fifth and a quarter of the world's untapped fossil fuel resources.

The United States also got involved in recent years when a team from the National Oceanic and Atmospheric Administration declared that warming had created a "New Arctic" that since 2000 has resulted in a decline in sea ice that was unprecedented in the past fifteen hundred years. The Arctic, they wrote, "shows no sign of returning to [the] reliably frozen region of past decades."

With $300 billion in potential projects underway, Russia remains the clear leader in developing Arctic infrastructure development. The world's largest country has moved to reopen some abandoned Soviet-era military installations and place new *facilities and airfields in its northern territory, while also establishing a string of seaports along its northern coastline. Ultimately, Russia hopes that offshore Arctic oil will account for as much as 30 percent of its production by 2050. Other nations with long-term interest in the region include Norway, Finland, Canada, and the United States. A good indicator of this activity is the use of icebreaker ships: Russia has fifty in operation; Finland, seven; Canada and Sweden, six apiece; and the United States, five. Scrambling to update its aging fleet, the U.S. Coast Guard plans to build six more (three heavy and three medium icebreakers), though the first won't be delivered until 2023.*[1]

While the pace of all the development is new, the most radical shift is the involvement of China—a nation with no territorial claim to the Arctic, just as it has also reached into Antarctica (discussed above). With its global economic and naval power on the rise, China has begun underwriting Arctic development projects despite its lack of territory there, underscoring the region's growing global importance. With a scientific research presence in the Arctic since 1999, China began the "Power of Siberia" project in 2019 as part of its effort to reduce its dependence on coal. This three-thousand-kilometer natural gas pipeline connects Russia's Siberian fields to northeast China.[2]

In 2013, Calgary, Alberta, the fourth largest city in Canada, experienced a significant natural disaster that disrupted its community, economy, infrastructure, and natural environment. Water from mountain-fed rivers, combined with saturated earth from a long winter and recent heavy rainfall, created a flood that inundated the downtown, the likes of which hadn't happened in a century. The heart of Calgary was closed as water drained from its streets. People and businesses were evacuated from their homes and livelihoods, and critical infrastructure impacts halted basic services. Instead of merely cleaning up, Calgary leaders designated the event as an opportunity. City leaders write:

> After experiencing our 2013 flood, we decided to be intentional about understanding the things that could disrupt our quality of life. The journey to develop this Resilient Calgary strategy has helped us identify what our greatest stresses and shocks might be, encouraged us to explore these disruptors and gather knowledge about our readiness for them, and compelled us to identify the specific goals and actions to move us forward toward our vision of Calgary as a sustainable, resilient place.[3]

Calgary is one of the world's designated "100 resilient cities." In our contested moment of energy transition, "resilient" has been used describe a new mode of urban planning that factors in the realities of climate change to assist cities in planning for sustainable futures. In planning their future, these resilient cities emphasize core strategies, including interconnecting environment and economic engines; having an openness to diversity and new peoples; learning from the past to consider our natural infrastructure as an asset when budgeting and planning; emphasizing sustainable and renewable infrastructure; and focusing on the future. In short, urban resilience is the capacity of individuals, institutions, businesses, and systems within a city to adapt, survive, and thrive no matter what kind of chronic stresses and acute shocks they might experience.

Calgary's efforts are particularly striking, of course, because it is one of the hotbeds of the early twenty-first century's other energy realities: extreme extraction that requires significant cost and technology to harvest remaining nontraditional sources. Its growth largely follows the boom-and-bust cycle of the energy industry. In the case of Alberta, the source of oil and gas is tar sands, discussed below. Energy companies headquartered here carry out one of the most intense extractive undertakings on Earth in 2020 and load sources of raw energy into pipelines and tankers to disperse it to the highest bidder all over the world. Despite this economic emphasis, Calgary envisions a future beyond fossil fuels for itself. Calgary has become particularly well known for its novel approach to funding resilient initiatives. For the last few years, Calgary has utilized a Resilience Dividend Tool that guides investment prioritization by quantifying the expected impacts from opportunities that make the city more resilient. A valuation of expected benefits helps make decision making easier by quantifying eventual benefits against immediate investment.

With this type of funding support, despite the fossil fuel industry's economic importance to Calgary, a major part of the city's plan is for a future organized by sustainable energy. First, the city approved a 2012 motion to purchase 100 percent renewable electricity to cover all electricity consumed in city operations. Next, in its Climate Resilience Strategy of 2018, the Calgary City Council established strategies to reduce contributions to climate change by improving energy management and reducing greenhouse gas emissions and to respond to a changing climate by implementing risk management measures to reduce the impact of extreme weather events and climatic changes on infrastructure and services. The Climate Resilience Strategy targets 80 percent reduction in community-wide emissions below 2005 levels by 2050, and aims to provide strategic oversight to climate-related activities in Calgary. The related initiatives now being implemented include a biodiesel pilot for vehicles, sustainable building practices, and the early adoption of LED technology (retrofitting more than eighty thousand public streetlights). The city's transportation network is also being enhanced with the addition of a new light rail system that will be completed by 2050. Whereas cities became a major focus of modernizing efforts in the developing world at the end of the twentieth century, now they have emerged as an important tool for implementing new ideas.

Examples such as Calgary and any of the other hundred resilient cities suggest that in 2020 an active energy transition is underway away from fossil fuel stocks. Applying what we have learned from previous transitions, environmental historian Christopher Jones offers another guiding lesson: transport matters, as do differences between systems; supply drives demand; energy transitions are overlapping and reinforcing; energy transitions create unequal geographies and reconfigure social power as well as mechanical power; and transitions are shaped by public and private actors.[4] With these ideas in mind, this chapter emphasizes more CURRENTS than have others for the simple reason that we exist in a fraught, exciting, and even contentious moment of energy use.

Starting from the same shared realization of energy centrality to human life and national success, a cacophony of constituencies vies to use this understanding to define our energy moment and, particularly, our energy future. In short, this chapter attempts to capture a moving target: our current energy transition. At its core, our current energy transition incorporates (albeit unevenly) the complex idea of energy that emerged during the twentieth century—starting from the dual realities that reliable energy supplies were a requirement of global economic and political power and that fossil fuel use imperiled our living environment—into plans for a more stable and sustainable human society.

To describe the volatile nature of this moment, picture an event that many of you may have been fortunate enough to experience: catching a large fish on a line. In an inevitable moment of elation and fury that is the only route to make any of it worthwhile, you must look down at the thrashing beast and figure out how to get your hands on it. How to control it. How to bring it into your realm. Staring at the beast that is the energy scene in 2021; for instance, one must find continuity between bicycle-share programs and fracking for natural gas, or between the Gulf oil spill and the proliferation of electric vehicles (EVs), or between local food initiatives and oil exploration in the Arctic. In short, we are witnessing and participating in an active energy transition—not every generation can say that! In order to capture the dynamic nature of our moment, this chapter extends the "CURRENT" and "CONDUIT" subheadings to include ". . . for Change" in order to suggest some of the patterns at work in our current transition. Look around to see what other currents and conduits you see moving humans away from our reliance on fossil fuels.

Here we go!

ENERGIEWENDE AS A CONCEPT FOR NATIONAL ENERGY DEVELOPMENT

On separate paths in the last few years, the environmental consciousness of the 1970s progressed to applied efforts of sustainable planning and design while the awareness of energy centrality pressed aggressive, extreme efforts to extract and to access remaining supplies of fossil fuels. At times, the two paths seem diametrically opposed, while at others they seemed to achieve a synchronicity that allowed observers to glimpse a different kind of human future. Most likely, our lifetime will be defined by these two paths, and a growing inevitability that the future that emerges will extend these moments of synchronicity to be the next great pattern in humans' life with energy. In other words, an energy transition.

Although many societies have implemented their own energy plans, a variety of German intellects since the 1970s have advanced the idea of *Energiewende* that seems to offer a clear middle ground that brings these two different developmental paths together in a sustained strategy. The concept begins with the recognition that energy policy and strategy require fundamental change but also that this change must come from restructuring the system to deliver equal levels of growth and prosperity based on already available technologies.[5] The true distinction of *Energiewende* is that it prioritizes that new system to free consumer choice from the limits of corporate and federal control. In the German language's unique ability to express nuance, *wende* means "a turnaround." Patrick Kupper writes, "People speak of an 'energy transition,' which is not quite the same thing: *transitions* happen, whereas a *Wende* is consciously pursued."[6] Felix Christian Mattes writes that *Energiewende* developed its strong support "not as a consequence of the rapidly escalating challenges, but primarily from its enormous optimism regarding the technological possibilities and the potential of new, decentralized, and especially consumer-oriented technologies and structures."[7] As an applied strategy, *Energiewende* coheres with and adjusts to the emerging information and science that today is so readily available to the consuming public. Instead of the lack of control over energy choices exhibited in developed nations since the 1970s, *Energiewende* is built around consumer control.

A significantly interdisciplinary approach, *Energiewende* disperses ideas and concepts broadly through the arts in order to influence consumers. For instance, the Deutsches Museum created a large-scale exhibition and book in 2016—*Energiewenden: Energy Transitions as Chance and Challenge in Our Time*—which eventually traveled the world. For any visitor, the necessity of energy in human life was put on display along with the complexities and challenges of supplying it. The shortcomings or limits of specific sources unfolded beside the pathbreaking innovations that enhance conservation or battery storage to extend the existing infrastructure.

In Germany, *Energiewenden* joined with a new generation of policy initiatives that had begun in the 1990s. These efforts culminated in decisions made at the G7 Summit at Schloss Elmau in 2015 that committed Germany to transition to an energy system without fossil fuels.[8] Mattes writes specifically of why the *Energiewenden*, since its inception in the 1980s, differs significantly from other initiatives when he writes that it "was able to materialize and develop such power and significance internationally because one country took the idea of a politically driven structural transformation of the energy system seriously and put a great deal of effort into implementing it."[9] Its success, however, relied on the inherent understanding of *Energiewenden* that technological innovation had to be supported by the cultural change lived and put into practice by consumers.

As a critical component to an energy transition, citizens' endorsement was not considered essential in *Energiewenden*; however, the widespread acceptance of such a shift among the populace was described with four requirements:

Orientation and understanding: explain the rationale and needs clearly;
Agency, or self-efficacy: provide citizens choice and empower them to adopt changes;
A net positive balance of risks and benefits: enumerate eventual benefits even while clearly communicating likely challenges;
Identification: create empowering mechanisms to allow citizens' actions to be clearly associated with the initiative.[10]

While Germany and a few other nations leaped forward to envision new energy possibilities, patterns of the twentieth century continued to play out elsewhere.

CURRENT: The Long Reach of the "Haves"

Calamity came around 1:00 a.m. on Monday, December 3, 1984, in a densely populated region in the city of Bhopal, Central India (which was briefly mentioned in chapter 6). The American company Union Carbide had located its pesticide plant in this populous location to facilitate hiring cheap laborers. India was a particularly hospitable location for chemical manufacturers as environmental regulations tightened elsewhere in the world after 1970. On this day, though, a poisonous vapor burst from the tall stacks at the plant leaked a toxic cloud of methyl isocyanate (MIC). Of the eight hundred thousand people living in Bhopal at the time, two thousand died immediately, and as many as three hundred thousand were injured. In addition, about seven thousand animals were injured, of which about one thousand were killed.

Industrial accidents might occur anywhere; however, the Bhopal incident represented a dangerous new era of quasi-industrial colonialism. In such a culture of economic development, was it right to refer to this incident as an accident, disaster, catastrophe, or crisis, or was it more appropriate to call it cultural sabotage, conspiracy, or massacre? In *The Bhopal Tragedy* William Bogard writes: "Each of these descriptions, in its own way, minimizes the problem of human agency and intention, and thus refuses to address directly the issue of responsibility."[11] For Bogard and others, Bhopal is a "tragedy" created by many dimensions of twentieth-century geopolitics (the relationship between geographical location, resources, and international diplomacy). Its connection to petroleum comes not from oil extraction but from the petrochemical industry, which by the late twentieth century made up a critical portion of humans' relationship with crude.

Petrochemicals, including consumer goods such as chemicals and plastics, dramatically altered the use of petroleum by offering a seemingly endless stream of applications. A selection of these chemical applications dramatically improved living conditions for many people in both the developed and less-developed world. Producing these chemicals, however, often created health hazards for surrounding communities. Fully aware of these outcomes, some manufacturers placed their factories in communities and nations that either would not complain or would not be noticed if they did so. In large and small examples, such planning serves as an example of environmental injustice or racism. Often, these human costs came only to create products organized by a system of perpetual need known as "planned obsolescence"—products designed for a limited life span that would thus require replacement. In the United States and many Western societies, this fed a sociological cycle known as "mass consumption," in which citizens' civic participation went beyond voting to include purchasing goods—feeding the economy.

The sociological shifts of this era of energy management possessed deep layers of irony for humans, particularly when petroleum reserves became increasingly difficult to locate. International corporations, the behemoths that had evolved from the oil companies described above as initiating an age of "petro-colonialism," exercised the ethic of extraction with a ruthless precision that dwarfed efforts of Standard and others to break into the jungles of Mexico and South America in the first decades of the twentieth century. In cases such as the Niger Delta, corporations such as Shell Oil worked out arrangements with political or military leaders that left residents with no financial benefit from oil exploration and all of its impacts on human health and regional ecology. Whether to harvest oil or to process it for consumption, entire societies were threatened to support the new formula that developed societies called "modern life." In addition to cities such as Bhopal, the ecology of oil crept into societies such as Nigeria that had little use for crude itself.

In short, while petroleum had woven deeply into the fabric of Western life after World War II, it both complicated and exacerbated the economic

gap between nations discussed by Marks above. For a century, many nations of high-energy consumption excelled and prospered. As a long-term
development strategy, conspicuously consuming crude oil, though, was
destined to eventually fail due to finite supplies. Even in the short term,
however, surprising outcomes of the addiction of the developed world
revealed themselves in distant locales and among people with little use for
the petroleum or the products it created. Resembling what ecologists call
feedback loops in the natural world, the outcomes of this phase in energy
use demonstrated that commodities needed by others could be deadly on
a physical and social level. While new, sustainable ideas about energy
emerged, the early twenty-first century remains defined by many aspects
of our life with fossil fuels. Even when many consumers know better and
wish for alternative options, the continuity of present-day life requires that
our life with fossil fuels continues—even as it becomes more and more
complex and distressing. In the following sections, this book explores
mechanisms for possible change in our current energy transition.

CURRENT for Change: The Emerging Science of Fossil Fuel Emissions

Beyond issues of supply, reliance on fossil fuels has also been complicated
by new scientific understanding of its environmental implications. For
most of the twentieth century, scientists worked to quantify the outcomes
of burning all fossil fuels, particularly petroleum and coal. By the second
half of the century, many humans in developed societies found that they
had a villain living among them: the internal combustion engine (ICE). In its
earliest version, this reimaging of the ICE had little to do with the growing
scarcity of petroleum supplies; instead, critics emphasized the inevitable
outcome of burning petroleum in car engines: emissions and air pollution.
Prior to the public understanding of concepts such as greenhouse gases
and climate change, air pollution was acknowledged to be unpleasant and,
very likely, unhealthy.

Air pollution, generated by industrialization in general, has been docu-
mented to possess health effects for humans since the early 1900s. The
connection between smog and auto exhaust is credited to Arie Haagen-
Smit, a German researcher at the California Institute of Technology. During
the 1950s, in order to bring his findings to the public, Haagen-Smit fought
off the savage criticism of the auto manufacturers, who claimed that a
well-tuned vehicle had no such adverse effects on the air.[12] Severe smog

episodes in California kept the issue in the public arena and helped to make it one of the primary issues for the nascent American environmental movement. The seminal event in the emergence of modern environmentalism, Earth Day 1970, contained many activities that related to air pollution.

The new appreciation of the environmental impact of the ICE was just the beginning of the problems that would face the brokers of America's high-energy existence. During the 1980s, scientists achieved even more detail regarding the environmental impact of auto emissions. Most important, the impact was not limited to local areas, such as the city of Los Angeles. New computer modeling combined with better understanding of the functioning of various layers of Earth's atmosphere to make clear that something was rapidly depleting the planet's protective ozone layers. In addition, heat was becoming trapped in Earth's atmosphere at an alarming rate, creating what has been called the greenhouse effect. Finally, by the 1990s scientists concluded that Earth was warming at a pace without historic precedent.

For most scientific observers, the root of each of these environmental problems—as well as others such as acid rain—was the burning of fossil fuels, which released massive amounts of carbon (in the form of carbon monoxide and dioxide) into Earth's atmosphere. The transportation sector alone is responsible for about one-third of our nation's total production of carbon dioxide. And, of course, the internal combustion engine is a primary contributor. By the end of the twentieth century, most observers accepted that not only was smog unpleasant and unhealthy, it may actually be contributing to the ruination of the entire Earth.[13]

Some scientists went even further. They argued that the burning of fossil fuels had broadened humans' environmental impact so severely that a new geological epoch should be named: the Anthropocene. Chemist Paul Crutzen argued in a 2000 article in *Science* that humans have become a geologic agent comparable to erosion and eruptions, and accordingly "it seems to us more than appropriate to emphasize the central role of mankind in geology and ecology by proposing to use the term 'anthropocene' for the current geological epoch."[14]

CONDUIT for Change: Who Saved the Electric Car?

Developed nations' reliance on mined energy resources created a century of inexpensive energy that has largely molded the society in which many of us now live. Most experts agree, though, that the next century demands a different model for its prime mover. Alternate sources of energy offer the possibility of reducing dependence on fossil fuels, which would also reduce American dependence on petroleum imports and also reduce pollution. Simultaneously, though, significant increases in energy efficiency and conservation measures would offer a great improvement and, in the short term, these adaptations might be the most easily implemented. For instance, a few simple ways to reduce energy consumption might include the use of lighter-weight automobiles and more efficient engines, improved house insulation, waste recycling, and improved public transport.

In terms of personal transportation, electricity emerged as the most likely replacement for ICE. The rise of environmental concerns focused on California in the late twentieth century and, therefore, it is not surprising that so did the development of electric vehicles. The California Air Resources Board (CARB) helped to stimulate CALSTART, a state-funded nonprofit consortium that functioned as the technical incubator for America's efforts to develop alternative-fuel automobiles during the 1990s. Focusing its efforts on the project that became known as the EV, this consortium faced auto manufacturers' onslaught almost single-handedly. Maintaining the technology during the mid-1990s, however, had been carried out by a variety of independent developers.

Absent government support and despite the contrary efforts of larger manufacturers after World War II, independent manufacturers continued to experiment with creating an electric vehicle that could operate cheaply and travel farther on a charge. The problems were similar to those faced by Edison and earlier tinkerers: reducing battery weight and increasing range of travel. Some of these companies were already in the auto business, including Kish Industries of Lansing, Michigan, a tooling supplier. In 1961, it advertised an electric vehicle with a clear, bubble roof known as the Nu-Klea Starlite. Priced at $3,950 without a radio or a heater, the car's mailing advertisements promised "a well-designed body and chassis using lead acid batteries to supply the motive energy, a serviceable range of 40 miles with speeds on the order of 40 miles an hour." By 1965, another letter from Nu-Klea told a different story: "We did a great deal of work on the electric car and spent a large amount of money to complete it, then ran out of funds, so it has been temporarily shelved." The Nu-Klea was not heard from again.

As mentioned above, in 1976 the U.S. Congress passed legislation supporting the research of electric and hybrid vehicles. Focused around a demonstration program of seventy-five hundred vehicles, the legislation was resisted by government and industry from the start. Battery technology was considered to be so lacking that even the demonstration fleet was unlikely. Developing this specific technology was the emphasis of the legislation in its final rendition. Historian David Kirsch writes that this contributed significantly to the initiative's failure. "Rather than considering the electric vehicle as part of the automotive transportation system and not necessarily a direct competitor of the gasoline car, the 1976 act sponsored a series of potentially valuable drop-in innovations." Such innovations would allow electric technology to catch up to gasoline, writes Kirsch. However, "given that the internal combustion engine had a sixty-year head start, the federal program was doomed to fail."[15]

The developments in electric vehicles that followed were mostly of a small-scale variety. The 1979–1980 Lectric Leopard from the U.S. Electricar Corporation of Athol, Massachusetts, was based on a Renault R-5 or Le Car, as it was known in the American market. One of the best-selling electrics was the CitiCar, built from 1974 to 1976 by the Sebring-Vanguard Company in Sebring, Florida. The CitiCar was essentially a golf cart equipped with horn, lights, turn signals, wipers, and an enclosed cabin with optional heater and radio. The CitiCar was succeeded by the Comuta-car, an identical two-seater. During the 1990s, a company known as Solectria built electric vehicles based on GM's Geo in Woburn, Massachusetts. Such efforts were celebrated by sustainable energy groups but remained well outside the mainstream of consumer vehicles.

It was just this distinction that seemed to make the EV1 different in the early 1990s. Developed with the support of the state agencies, the EV was then leased to consumers by GM in California and Arizona in the late 1990s. There were significant costs involved, because customers were required to have home-charging stations to keep their EV1s fueled. GM worked with the state to establish a few charging stations in shopping centers and office buildings. Despite what seemed like a significant example of industry and government cooperation, the EV did not turn out well. So revealing as an episode of a missed technological transition, the EV1 became the subject of the well-known documentary film: *Who Killed the Electric Car?* As the title belies, although the car gained a passionate following among some users, the project died when California backed down on its mandate for zero-emission vehicles. The primary reason given for EV's demise was difficulty developing battery technology. There appears to be more to the story, though.

Journalist Jim Motavalli describes the fight over the EV1 as a California battlefield in the mid-1990s. From billboards to radio talk shows, auto companies spent approximately $34 million to depict CARB as an extremist political group who wished to force Americans from their right to choose their automobile. Thus, just as some of the auto companies were developing electric vehicles that would satisfy California's guidelines, they were also orchestrating a publicity campaign to do away with the agency overseeing the guidelines. *Car and Driver* magazine called CARB "the most environmentally draconian government agency in the nation."[16] GM withdrew support for the two-seat coupes. The cars were ordered back to the manufacturer for crushing, much to the dismay of a vocal group that fought to keep them on the road. GM was not the only automaker to cancel production of battery-electric vehicles during the 1990s so it could focus on hybrids and hydrogen fuel cells; with each cancellation, the chances of success for a mass-production of battery-electric cars seemed less likely. Clearly, the transportation market was an open battlefield for innovation; however, ICE wasn't going to be given up without a fight.

CONDUIT for Change: Giving Diesel and Biofuels Another Look for Transportation

Although they were resisted by large carmakers, after the 1970s, many of the initiatives for alternative fuels took shape in laboratories ranging from massive agricultural and petrochemical corporations to institutes of higher education to, of course, the garages of private citizens. In fact, experiments with what have become known as "biofuels" have never really stopped. After an early history of success, biofuels largely lost out to the use of inexpensive fossil fuels in the expanding U.S. economy. In many European nations, national governments sponsored experiments with various forms of biofuel development. In the United States, the federal government had less involvement. American experiments were primarily carried out by agricultural experiments and particularly emphasized the use of soybeans. By the end of the twentieth century, though, a variety of other experiments had progressed.

From the 1970s to the present, European nations were much more keen to emphasize consumer vehicles powered by diesel. Many European governments encouraged drivers to buy diesel cars as an alternative to traditional gasoline-powered vehicles because they were more efficient. Thanks to tax breaks and other incentives, diesel cars now make up about a third of all vehicles in Europe. As scientists learned more about auto emissions, diesel proved to be a dangerous source of additional pollutants, particularly soot, particulates, and nitrogen oxides (NOx). In an ironic twist, the European diesel initiative that grew from an interest in using less petroleum has today resulted in more concentrated pollution in many cities. Whereas in 1990, just 10 percent of new car registrations in Europe had run on diesel, by 2011 that had climbed to nearly 60 percent.[17]

Most large-scale experiments with biofuels focused on ethanol, which was discussed above. One of the most significant undertakings grew from the efforts of Ag Processing Inc. (AGP). In 1994, AGP, the country's largest soybean-processing cooperative, formed a new joint venture called Ag Environmental Products (AEP). In 1996 AGP opened a new batch-process biodiesel plant with a capacity of five million gallons (17,500 tons) in Sergeant Bluff, Iowa, adjacent to an existing seed-crushing facility. Such an operation represented an exciting, new frontier for biofuels.

In the spring of the following year AEP provided biodiesel fueling stations at ten farm co-op locations in six Midwestern states. Other stations were subsequently added. Over the years the $6 million soy methyl ester facility at Sergeant Bluff has produced a wide range of products, including biodiesel, solvents, and agricultural chemical enhancers under the SoyGold brand name, which were promoted and marketed for AEP. Biodiesel produced by AGP has been used in a wide range of vehicles by customers across the country over the last few decades.

Other, similar operations were established throughout the world after the 1990s. These experiments set the stage for an exciting new era in energy when the high price of petroleum again made it feasible to introduce alternative fuels.

CURRENT for Change: The "Radical" Science of Climate Change Becomes Mainstream

The acceptance of fossil fuel emissions' role in creating smog and other issues of pollution marked an important step on the way to humans making more complex connections. Primary among these, the scientific understanding of climate change emerged as an international undertaking at the end of the twentieth century. In 1985, the Vienna Convention for the Protection of the Ozone Layer argued that ozone depletion demanded action or it would lead to numerous health problems (such as skin cancer) brought on by exposure to UV radiation, which would eventually cost society substantial sums. Two years later, in 1987, the Vienna framework was given teeth in the Montreal Protocol. The protocol specified formal emission restrictions and led twenty nations (including the United States) to sign on. This remarkable international agreement provided climate scientists, politicians, and members of the general public who were increasingly concerned about global warming with a model for action. As a consensus took shape in the science behind global warming, the call for action increased.[18]

In the United States, a number of scientific reviews sounded increasingly strong warnings to policy makers. Arguably, the rise of global warming as a national and international political issue exemplifies the ever-greater development of public knowledge and democracy—simply, the public was much more aware of basic ecological principles than at any time in the past. By the late 1980s, as the result of increased awareness of environmental problems, greater reporting in the media, and responsiveness of the political system, global warming was on the cusp of receiving a place in the public consciousness and was ready to take off—should something happen to demand action.

The year 1988 came as if on cue. First, the heat waves and droughts of the American summer of 1988, especially in the central and eastern states, made a great deal of news in magazines, newspapers, and TV programs. The National Climatic Data Center (of NOAA) estimated that there were about $40 billion in economic losses and five thousand to ten thousand deaths. At the start of this difficult summer, in late June, NASA scientist James Hansen gave testimony to Congress in which he said that it was virtually certain that human activity was responsible for the global warming trends and that this might bring more storms, floods, and heat waves in the future. Hansen made the now-famous remark: "It's time to stop waffling . . . and say that the greenhouse effect is here and is affecting our climate now."[19] Polls indicated that, over the next year, public awareness of global warming rose considerably. However, the United States possessed political and cultural baggage that would not allow it to lead global action on this issue.

Primary leadership on climate change action came from the United Kingdom, which possessed a particularly strong interest in the issue because, as a relatively small nation geographically, it stood to lose a great deal from climate change. For one thing, its temperate climate was dependent on the Gulf Stream, which was potentially endangered by global warming. In 1988, the conservative prime minister Margaret Thatcher (who had a degree in chemistry from Oxford University) became the first major politician to embrace global warming as an important issue. After being briefed on the science of global warming by her adviser Crispin Tickell, Thatcher brought the issue to the attention of the nation's scientists in a major speech to the Royal Society.[20]

In 1988, rising scientific interest and political concern about global warming gave rise to the foundation of an international agency, the

Intergovernmental Panel on Climate Change (IPCC), by the World Meteorological Organization and the United Nations. The IPCC's task was to organize the work of an international group of scientists to produce periodic reports summarizing the best knowledge on the world's climate in order to help political leaders make policy decisions. U.S. president Ronald Reagan was strongly in favor of forming the body. Through a series of reports starting in 1990, the IPCC began a decade-long process of attempting to explain the implications of climate change to the world community. Throughout the 1990s and into the twenty-first century, the consensus on anthropogenic global warming continued to deepen. A remarkable new reason for relying less on petroleum was emerging before humans' eyes at the dawn of the twenty-first century.

From these scientific understandings, a steady push grew for international mitigation of the factors thought to contribute to climate change. In 1997, the UN organized the Conference on Climate Change in Kyoto, Japan, which ultimately resulted in the Kyoto Protocol, fully ratified in 2005. Moving the policy and scientific reality to the mainstream, Al Gore and film producer Laurie David created *An Inconvenient Truth* to serve as a counterbalance to the misinformation about global warming that they felt was being sent to the public through popular and political culture. By designing a film that would receive a large release and by teaming it with other media outlets, including a book authored by Gore, Gore and David brought new energy and awareness to the issue, which culminated with an Emmy Award and a Nobel Peace Prize in 2007 for Gore and the IPCC.

Although climate change remained a volatile political issue in many developed nations, the initiative of the ensuing IPCC fueled international cooperation in the formation of COP-21 in 2014–2015. Primarily emphasizing aspirational goals, this international accord joined over two hundred nations in efforts to limit emissions in the early decades of the twenty-first century. It was unified around the purpose of holding the increase in global average temperature to well below 2°C above preindustrial levels and to ensure that efforts are pursued to limit the temperature increase to 1.5°C. To achieve this, the Paris Agreement stipulated that all countries shall review their contributions to reducing greenhouse gas emissions every five years. Each new contribution set out on a national level should include a progression compared with the precedent. The parties committed to reaching a global peak in greenhouse gas emissions as soon as possible, in order to achieve a balance between emissions and their removal in the second half of the century. Nations were also required to increase their efforts to mitigate and reduce their greenhouse gas emissions. Historic as a demonstration of the global acceptance of the scientific reality of climate change,

COP-21 lacked "teeth" and remained susceptible to changing political winds in each nation, particularly in the United States.

As one of the primary contributors to potentially damaging emissions, the burning of petroleum swiftly became part of the new equation that strove to factor in climate change as a consideration for development strategies. Although a sweeping shift from ICE has not yet taken place, climate change offered a new logic to fuel global interest in an energy transition away from petroleum's use, particularly for transportation.[21]

CURRENT for Change: Hubbert's Peak and the Reality of Extreme Energy Supplies

Emerging out of the 1970s, the finite supply of fossil fuels such as petroleum became much more predictable with the help of mapping technologies; however, geologists had long argued its finite existence. Based on the general theories of petroleum geologist M. King Hubbert, "peak oil" ran contrary to the culture of petroleum in nations such as the United States during the twentieth century. The reality of petroleum has always been that it will run out, even when suppliers tried to convince us otherwise, whether at Spindletop in Texas, in Bahrain, or in Saudi Arabia. Today, acceptance of this reality is referred to as "peak oil."

When Hubbert, who was working for Shell, first forecast in 1949 the brevity of the petroleum age, his employer and many professionals called him the latest in a century's worth of "Chicken Littles"—skeptics predicting the impending end of petroleum supplies. In 1956, Hubbert put a point on his argument by focusing on the American domestic reserves, which he forecast would peak within thirty to thirty-five years and then slowly decline. His professional standing did not change until his forecast proved accurate, when American production reached its peak in 1970. His theory became known as "Hubbert's Peak," and geophysicists set out to apply his calculations to the known global supplies. Kenneth Deffeyes, for instance, reports that this global peak occurred in the first decade of the twenty-first century.[22]

In the intervening years, petroleum geology changed dramatically. Seismic mapping now made it possible to map quite accurately the petroleum reserves that lay untapped beneath the Earth's crust. With this additional technology, estimates of reserves—and theories such as Hubbert's—gained considerable credibility. Within this accepted paradigm, the primary variability became how increased competition, particularly as

India and China industrialized, might make the supply's demise come even more rapidly than forecast. Although it remains nearly impossible for geologists to focus on a specific date and some critics continue to quibble with Hubbert's computations, by the early twenty-first century energy forecasters began to change the culture of petroleum to reflect an awareness of petroleum's impending decline in supply. Large international oil companies began to diversify their efforts somewhat, particularly in the public sector. For instance, in 2001 British Petroleum (BP) actually changed its official name to BP and its slogan to "Beyond Petroleum."

Whether or not the industry for which Hubbert worked openly adopts his theory, the corporate culture of Big Oil radically changed in the twenty-first century. The prudence of relying on endless supplies of crude had been called into question by geological and ecological reality. To negotiate these new realities, energy companies flexed their approach. For instance, when peak oil forecasted thirty or fifty additional years of oil, it based calculations on traditional reserves. In the first decade of the twenty-first century, though, energy companies adopted more costly efforts to develop nontraditional reserves that fooled the peak oil forecasts—at least temporarily.

New methods of drilling and development allowed American companies to develop less traditional reserves, including the Bakken Formation in North Dakota and others in Texas and Alberta, Canada. In particular, horizontal drilling and hydrofracturing have transformed these regions and others into prolific producers of oil and natural gas. In this moment of reconfiguration and desperation, nations such as Russia have adopted more nontraditional methods, including political manipulation of various sorts, and North America has become one of the world's largest oil producers. Often, oil companies argue that this spike in production undercuts alarmist ideas such as peak oil; instead, though, these methods are better seen as the industry's last gasp. Together, all of these nonconventional methods to access and develop energy resources demonstrate the intensifying importance of the remaining supplies worldwide and the extreme measures that have now become business as usual for acquiring our power. The following sections cover a series of examples of methods that are based in this extreme energy reality.

EXTREME CURRENT: Gulf Spill Live Video Feed, 2010

As the camera pans between Lady Gaga and Larry King, at the bottom of the screen CNN made sure to provide viewers with not just the latest scores and stock numbers but also a live, inset image (approximately 1/10th of the screen) of the Deepwater Horizon wellhead leaking beneath the Gulf of Mexico. Not to be outdone, each competing news channel soon included the live video feed of the sheared-off industrial pipe releasing black and gray plumes. A churning cloud of impenetrable grime, the plumes, we knew with neither explanation from technological experts nor corporate leaders, were oil and natural gas. Once it passed beyond the limits of our live video feed, its destination would be the marine ecosystem of the Gulf of Mexico, bathing coastal birds in a coating of crude and introducing petroleum into the deepwater ecology.

With neither narrative nor drama, the live video feed, never deviating from its constricted focus on the wellhead, merely suggested these larger implications of the BP oil spill of 2010. In a related narrative, many of us followed the surface implications through other news reports. Normally, this would be our only view of such events. The live video feed provided a unique continuity to the story of the BP oil spill—cause and effect. In the viewfinder, illegible words and numbers were the only additional details one could observe, giving the live video feed a suggestion of scientific veracity— as if generated by a laboratory's measuring device. Or, to many viewers, the numbers and gauge readings provided the appearance of outer space or a scene from a science fiction film. Its main purpose was exactly the opposite, though; the live video feed depicted the spill's veracity and proximity.

The Gulf spill of 2010 was an unprecedented ecological disaster worsened by the lack of an organized response. Although problems with the response to the spill and the disturbance of the Gulf's delicate ecosystem drew similarities between it and Hurricane Katrina, the spill also possessed the implications of an unnatural disaster. In particular, the spill very publicly illustrated the failings of a corporate and regulative culture structured by the need for cheap energy.[23] Following the drilling rig accident that killed eleven workers, the response often complicated this reality. Dispersants, intended to break up the crude, were released deep in the ocean and are now feared to have further polluted the Gulf. The region's fishing industry, the nation's largest, was entirely shut down for months. A moratorium was placed on further deepwater oil drilling. For nearly three months, hapless corporate and political leaders poised themselves for a quick end to the leak and spill; instead, the oil and gas continued to gush into the Gulf for more than a hundred days. When the well stopped gushing in late July, the live video feed showed viewers that, too. In total, the experience of this spill provided global viewers with a clear example of the potential costs of reliance on petroleum, particularly in extreme environments such as the deep ocean.

The live video feed arrived on viewers' television or computer screens through the use of a series of complex undersea tools. Just as undersea robotic devices carried out much of the repair work on the wellhead, each device also contained cameras. Remotely operated vehicles (ROVs), normally owned by one of the large technical supply companies and leased by BP or another oil giant, carried out operational and maintenance duties by providing a visual link to the undersea activities that have come to dominate current oil exploration. ROVs and the infrastructure used to harvest petroleum from deep in the ocean are an indication of great technological advancement; however, they also indicate the growing scarcity of petroleum, which now draws humans to more complex environments that require greater expense. Before they provided 24-7 coverage of spillage, ROVs used their down time to increase our knowledge of the marine world, from locating lost shipwrecks to charting the migratory pattern of the Greenland shark.[24]

The White House joined Rep. Edward Markey, a Massachusetts Democrat, to demand that the Deepwater Horizon leak be shown to viewers live, as it happened. "This may be BP's footage, but it's America's ocean," Markey said in a statement. "Now anyone will be able to see the real-time effects the BP spill is having on our ocean. This footage will aid analysis by independent scientists blocked by BP from coming to see the spill."[25] Almost immediately, private scientists joined with NOAA and other federal experts to protest BP's low estimate of flow rates. BP initially put the figure at 1,000–5,000 barrels per day (bpd)—or no more than 210,000 gallons. By July 2010, most experts agreed that all along the flow rate had been 25,000–80,000 bpd. Analysis of the size and dimensions of the plume in the live video feed offered a primary source of external evaluation.

EXTREME CONDUIT: Natural Gas Fracking and Debate

While oil producers did not necessarily accept forecasts of peak oil by Hubbert and others, their willingness to drill in the deep ocean was one indication that the prudence of relying on endless supplies of crude had been called into question by geological and ecological reality. To negotiate these new realities, energy companies have flexed their approach. For instance, when peak oil forecasted thirty or fifty additional years of oil, it based calculations on traditional reserves. Mentioned above, though, in the first decade of the twenty-first century, energy companies manufactured new methods to develop less-traditional energy reserves, including the pursuit of underground reserves through hydrofracturing.

Referred to as "fracking," this form of extraction proved successful in Texas starting in 2008. Underground seams of shale could be accessed through horizontal drilling and then injected with hydraulic fluid at strong pressure. This process of fracturing the shale releases natural gas that can be captured as it gathers in the underground fissure. Although technically challenging and expensive, fracking created decades of boom in many areas that were not currently under energy development. For instance, the Marcellus Shale Formation that extends in the United States under Pennsylvania is largely located beneath agricultural lands. Since 2010, energy development companies have moved into the region from Texas and elsewhere to develop the "play" while it remains cost-effective. But natural gas is challenging to move and the effort to make it profitable has made developing this energy source even more complicated.[26]

EXTREME CURRENT: Putting Extreme Energy to Work in Monaca, Pennsylvania

The effort to put the extreme natural gas gotten by fracking to work has required the construction of a complex labyrinth of pipelines throughout the mid-Atlantic in recent years. But pipelines to carry natural gas are complicated and expensive to build. For this reason, the industry's dream has been to construct nearby industries that could put the energy directly to work. To that end, in 2020 Monaca (pronounced mu-NAA-ka), Pennsylvania, bustles with development and construction. A large employer has moved into the area with the promise of hiring thousands, and the community prepares for ancillary possibilities. The hub of action is a large industrial plant that will draw from the abundant, inexpensive energy that is available from western Pennsylvania's reserves of fossil fuels. The particular draw for construction at this moment is newly discovered sources of energy that will make the factory's work less expensive to pursue in this locale than in most others.

This scenario, of course, could have been playing out in the nineteenth century, when coal supplies made feasible the plants that would undergird America's path through industrialization. The blend of available raw materials and cheap energy helped to make Pittsburgh the vital hub of the era and headquarters for great corporate entities such as US Steel. Instead, however, Monaca's opportunity unfolds today as one of the world's largest corporate entities, Royal Dutch Shell Oil, seeks to take advantage of the natural gas acquired through hydraulic fracturing throughout the state of Pennsylvania. Once again, in this twenty-first-century model of industrialization, available energy supplies attract applications. In the Monaca project of 2019–2020, for the first time in two generations industrial infrastructure is not being dismantled; instead, new parts are being assembled by Bechtel, Shell's primary contractor, into a world-scale, state-of-the-art chemical processor—known as a "cracker plant"—to convert liquid natural gas into polyethylene, a common plastic. The Monaca factory is an indicator of a new era in Americans' use of fossil fuels that is defined at least partly by its continuity with past patterns of industrialization.[27]

Even a site such as Monaca must be seen within the energy transition in which it participates. Today, we think differently about our entire energy system. Infrastructure such as powerlines and batteries can become important conservation mechanisms if they function more effectively. And burning fossil fuels for tasks such as personal transportation or to manufacture temporary objects meant for single usage appears to be a wasteful source of a significant amount of climate-influencing emissions. Our energy accounting has changed, particularly as we observe increases in the outcomes of climate change, including increased temperatures and erratic weather patterns, melting ice caps, rising seas, and the heightened intensity and frequency of storms. Monaca may provide momentary economic boom for a region; however, it is based on a previous, outmoded model of growth whose energy—derived from "extreme," temporary supplies—will not last.

EXTREME CONDUIT: Pipelines, Alberta Tar Sands, and Cyberenergy Warfare

Similar to fracking, other forms of extreme energy have also been defined by complexities of development. In one of the clearest examples, high oil prices made it cost-effective in Alberta, Canada, to remove extensive forests and layers of Earth to access thick, tar sands that occur beneath the surface. Once accessed, the tar sands are moved through a conversion process that requires significant energy input in order to produce a usable petroleum

product. While this is a challenging and expensive process, the true difficulty has derived from the remoteness of the product. To connect these tar
sand development sites to markets requires massive pipeline construction
through challenging terrain over very long distances.

Therefore, in order to take advantage of the energy that can be harvested
from Alberta tar sands, the industry has proposed and begun carrying out
the construction of more than ten thousand miles of pipeline at an estimated cost in excess of $40 billion. In one direction, pipelines cross the
relatively short distance of British Columbia to deliver the manufactured
crude oil to Kitimat and Burnaby, ports that can then load tankers for shipment all over the world—with a particular emphasis on China and other
nations in Asia. The much longer and more complex undertaking, known
as the Keystone XL pipeline, spans a much longer distance to bring oil
through the American South. In particular, the Keystone XL is intended for
the Houston, Texas, region where the oil can be processed into a variety of
petroleum products and then shipped elsewhere.

In the waning days of our reliance on fossil fuels, such extreme measures
have also revealed that the complex enterprise of developing resources
such as the tar sands may simply be too risky and costly—too extreme. As
the political winds changed in the United States in 2021, the Biden administration rejected further development of the Keystone XL pipeline, and the
corporation pursuing it declared for bankruptcy after facing extensive legal
challenges. Simultaneously, Russian cyberterrorists identified the tenuous
connection of the industry to its computer-run pipeline infrastructure and
forced the temporary closure of the American Colonial pipeline with a
"ransomware attack." On seemingly every front, the call from critics of
the fossil fuel industry to "leave it in the ground" seemed to gather new
credibility.

FACTORING GREEN INTO OUR ENERGY NEEDS
BY "LEARNING TO LIVE WITH LIMITS"

Although the energy decisions carried out by industrial entities took on a much
more intensive approach after the late 1900s, a clear alternative approach grew
out of the environmental thought of the 1970s. Eventually, this green culture
allowed many consumers in the twenty-first century to inform energy choices
with more of a sustainable ethic. This market shift began in the 1970s as the
energy crisis stimulated interest in alternatives and also cast a clearer view of the
dependence of developed societies on volatile and unreliable petroleum supplies.

In terms of intellectual shifting, the reality of petroleum dependence had begun to emerge in many ways by the late 1960s. Some of these realizations grew from the advance wisps of new scientific understanding, ranging from oil spills to acid rain. Others grew from the wake of a complex social movement to reconceive American patterns of consumption. In *The Genius of Earth Day*, environmental historian Adam Rome argues that the greening culture that took shape during the 1970s created a "new eco-infrastructure." Ranging from "e-beats" for journalists to "eco-books," Rome writes: "The post–Earth Day eco-infrastructure gave the environmental movement staying power. After the passions of Earth Day cooled, the first generation of environmental lobbyists ensured that politicians still felt pressure to protect the environment."[28] Together, influences ranging from journalism to politics and from education to planning helped to create the critical, shifting terrain of consumption after the 1970s that affected many aspects of American life, including energy consumption. Emergent green thinking proved to be a crucial catalyst for the energy transition from petroleum dependence. Americans after the 1970s were forced to confront the ethics that drove their ecology of oil; however, culture-wide ethics change very slowly.

Most often, historians boil these socioeconomic factors of the 1970s into the term *Arab oil embargo* to denote the hinge-point of change in the dynamics of American energy supplies—not of consumption patterns.[29] This proves to be a dangerous oversimplification when one attempts to trace each strand of these patterns. In fact, the actual event when OAPEC cut its oil shipments to the West in the 1970s is merely one formative moment—albeit critical—in a decade-long remaking of the way that American consumers viewed their petroleum supply. The distinctions between our views on the importance of the oil shock may reveal more about our understanding of certain terms than about whether any change can be measured. If we properly interpret the term *energy transition*, for instance, and do not expect a "crisis" with immediate results, the American reaction to the 1970s oil shock emerges as a seminal hinge in consumers' interpretation of the nature of energy supplies. In fact, the gradual nature of this change also reveals a critical dimension of the 1970s oil shock: the application of a new environmental perspective into everyday life and also into regulative policy.

More broadly, though, these implications are not just about transportation; they also suggest broader changes in ideas of energy.[30] Termed "learning to live with limits," a socioeconomic trend emerged in the 1970s that forced hard realities on to the American consumer at every turn.[31] For a generation that knew the expansive culture of energy decadence from previous decades, envisioning scarcity felt like a national failure. More than any other resource, petroleum taught 1970s Americans that times had changed. With the absence of excess supply, our culture of petroleum seemed ridiculously disconnected from reality. Our fetish with size and power in vehicles, for a fleeting moment, seemed juvenile. Although the ebb and flow of the public's zeal for conservation might be found in other modern societies, the United States, consuming as much as three-quarters of the world's petroleum supply at times in the twentieth century, presented the greatest challenge for the implementation of even a commonsense paradigm of energy management or regulation.[32] In particular, these cultural impulses met in the 1970s as American consumers and their political leaders reconsidered the concept of conservation and began to reconfigure it as a primary component of future technological innovation.[33] Taking shape in the twenty-first century, this sustainable ethic took the form of a market for alternative energy production and use that became more and more available to green consumers.

Elon Musk Defines a Growing Market for Green Energy Products

While some entire nations identified alternative energy strategies to make a more sustainable future for themselves, Elon Musk, South African inventor and innovator, emphasized products that could be bought by individuals as part of the green marketplace. Most important, his electric vehicle company, Tesla, became the first manufacturer to combine consumers' interest in a luxury-level automobile solely relying on a cutting-edge electric engine. The first and only exclusive manufacturer of electric autos, Teslas are not sold through dealers; instead, in many high-end shopping malls, kiosks allow customers to custom design their vehicle and place an order for a Tesla that will be delivered to them. While Tesla sold its one millionth car in 2020, new programs in China and elsewhere suggest 2021 might be the best year yet.[34] In many countries, the company has stimulated sales by also investing in infrastructure such as charging stations and roads made from material to charge the vehicles. In 2019, Tesla became Norway's best-selling vehicle brand—of any type. And, the government is striving to have every Norwegian vehicle powered by electricity by 2025!

Building on this foundation, though, Musk solidified the high-end-consumer marketplace as an emerging growth sector. These consumers, Musk correctly bet, were willing to spend more in order to purchase energy-related items and systems that cohered with their commitment to a sustainable ethic. Here are just a few examples:

- Called the Powerwall, Tesla currently manufacturers a battery storage system that is intended to be integrated into homes, similar to a furnace or air-conditioning system. Particularly with homes using solar and wind collection, the Powerwall allows homeowners to retain their collected power instead of sending it to the community's overall energy grid.
- Similarly, the Tesla roof is the most advanced version of solar-collecting shingles that entirely replace a home's roof with a system for collecting energy from the Sun.

While each innovation suggests an exciting energy future, more than anything Tesla's efforts suggest the clear existence of a green economy for consumers willing to spend a bit more in order to ensure that they live as sustainably as possible. Tesla has made sustainable energy more available while also helping to carve out a new green economy that can succeed in the free marketplace.

Dieselgate Reveals the Power of Green Consumption

The best indicator of the power of the emerging green marketplace might be the effort of corporations to leverage the interest in sustainability to mislead consumers, which is referred to as "greenwashing." Most observers of the general news cycle from 2016–2017 possessed at least a glancing knowledge of the corporate controversy that transpired from the release of this new, clean diesel "technology." In fact, though, in his expose *Faster, Higher, Farther: The Volkswagen Scandal*, Jack Ewing describes the green diesel initiative as a well-thought-out business initiative by Volkswagen/Audi during the first decade of the twenty-first century. The initiative was so well thought out that it distinguished between customers in Europe and those in the United States.

"In Europe," writes Ewing, "with its astronomical fuel prices, Volkswagen could market diesel on fuel economy. In the United States, where gasoline was cheaper than diesel and much less expensive than in Europe, Volkswagen needed another pitch." In particular, Volkswagen sought a wedge that would allow it to specifically challenge Toyota in the United States, the company that it most identified as its market rival. Specifically, VW focused on the remarkable success of the Prius to "lend its owners a green halo."

Shot to mimic the cliché-feel of a TV cop drama, the 2010 advertisement for Audi/Volkswagen vehicles shows cars waiting at a road block "eco-check." Impossible to view without irony today, the Green Police are inspecting each vehicle's efficiency. An officer in green shorts spies the driver of an Audi A3 station wagon and calls out: "Got a TDI here—clean diesel. You're good to go, sir." And the Audi is guided out of the line of vehicles and allowed to pass the non-TDI vehicles in the queue.[35]

Although VW could not compete in hybrid technology, it sought to market "clean diesel" as an update on the pitch that presented the Prius and other hybrids as "a badge of antimaterialism and a form of protest against the ostentatious gas guzzlers churned out by [American manufacturers]."[36] The campaign began in 2008 and focused on the new sedan Passat, and by 2011 it had resulted in a significant increase in U.S. sales. VW ran into difficulty when it sought to expand the market further and to specifically seek to grow American consumer interest in diesel vehicles. In order to accomplish the desired "pitch," VW employed software that would disguise emissions problems and to falsely enhance mileage statistics for its diesel engines. Although the vehicles were quite efficient,

VW sought to tip the regulative scales in order to make the engines nearly on par with hybrids.[37]

Remarkably, when independent testing revealed this deception in 2013, VW responded with one of the most blatant cover-ups in recent corporate history.[38] The total scandal, then, resulted in a $15 billion settlement, an array of lawsuits, and near total collapse in the VW brand. The place of the VW scandal in corporate history continues to play out; however, it is not too early to assess its significance as part of humans' use of energy for transportation. In the history of humans' use of automobiles, the VW scandal—possibly more than general market trends in the sales of hybrid and electric vehicles—reveals a dramatic cultural shift in consumers' acceptance of thinking green about their vehicle choices. In short, a successful global manufacturer such as VW would not choose to "greenwash" unless it also perceived a viable market shift that it wished to exploit.

CONCLUSION: MODELING TRANSPORTATION ALTERNATIVES ON THE GLOBAL STAGE

The market share that VW pursued in "Dieselgate" had much more success with the rapid development of electric and hybrid vehicles after 2000. Similar to during the early twentieth century, in terms of emerging technology the transportation sector is wide open for new possibilities in 2021. Although Toyota broke through the frontier of reliable hybrid vehicles, its safety problems uncovered throughout 2010 broke its juggernaut on the automobile market. New technologies are being developed by new manufacturers all over the world in what promises to be a dynamic new era in human movement, somewhat similar to that witnessed a century ago. One of the most significant differences in this era of innovation, though, is its truly global nature. The development of early autos was a transnational undertaking; however, it was entirely driven by European and American inventors. Today's manufacturers are often much more than inventors, and the markets that they seek to develop extend through Southeast Asia to emphasize India and China.

In one of the most remarkable experiments, over the last few decades Brazil's leaders have used their quasi-dictatorial control to impose some of the most radical alterations in the transportation sector. In 2007, after securing enough fuel crops in production, Brazil committed itself almost entirely to the use of biofuels for transportation. Thirty years prior, the nation, which had imported 75 percent of its oil, reacted to the 1970s oil crisis by setting out to use ethanol made from sugarcane to liberate itself from imported oil. When the OPEC oil embargo crippled the nation's economy, Brazil's dictator at the time heavily subsidized and financed new ethanol plants. In addition, he directed the state-owned oil company, Petrobras, to install ethanol tanks and pumps around the country. With his centralized political control, he offered tax incentives to Brazilian car-makers to encourage them to crank out cars designed to burn straight ethanol. When democracy was restored after 1985, Brazil's centralized initiatives provided a head start for an energy transition.

The conversion of Brazil's transportation sector has established it as an example for other nations. By the mid-1980s, nearly all the cars sold in Brazil ran exclusively on *álcool*, or ethanol. Just a decade later, the ethanol industry had become self-sufficient and no longer required government subsidies. "Today," writes journalist Joel K. Bourne Jr., "nearly 85 percent of cars sold in Brazil are flex: small, sporty designs that zip around the lumbering, diesel-belching trucks in São Paulo." The Brazilian vehicles do not rely on corn-based ethanol, which uses only the kernel and requires the use of enzymes to break down starches and begin the fermentation process. Instead, Brazil's method uses the entire sugar-cane stalk, which is already 20 percent sugar and begins fermenting almost as soon as it is cut from the field. Industry leaders estimate that sugar-based ethanol creates 55–90 percent less carbon dioxide than gasoline.[39]

The ability of China to so strictly control its path toward industrialization, of course, relies on strict centralized authority. For instance, in a second example, China, expediting its movement toward industrialization in the second decade of the twenty-first century, has become one of the world's largest consumers of fossil fuels; however, it does so while also emphasizing renewable sources in both energy production and transportation. Thus, when the global recession shrank petroleum demand worldwide, China swept past the United States to become the largest importer of oil from Saudi Arabia.[40] In addition to constructing re-fineries in China specifically to process Saudi crude, China is constructing two refineries in Saudi Arabia for this purpose. China's rapid energy growth, however, is based mainly in coal, which supplies almost 80 percent of the nation's energy. As part of its centrally controlled drive toward industrialization, though, China has designed a low-carbon path decreasing its energy intensity—energy used per unit of economic output—by 20 percent in coming decades.

The centerpiece of this low-carbon industrialization strategy began with China's Energy Conservation Law, formed in 1997. Updating the law in October 2007, the National People's Congress made it one of few in the world requir-ing practical implementation to promote comprehensive energy conservation and providing the legal basis for long-term resource conservation in China. The Energy Conservation Law now reads:

> Conservation of resources is one of China's basic national policies. The national energy development strategy is to implement both energy conservation and energy development simultaneously, while putting energy conservation first. . . . The State

Council and local people's governments above the county level should integrate energy conservation into the national/local economic and social development plan and the annual development plan, and should coordinate the preparation and implementation of specific annual and long-term energy conservation plans. . . . Energy conservation target obligation and assessment will be implemented by the state, and the completion of energy conservation goals will be used as appraisal factors for local governments and for all parties responsible. . . . The state implements industrial policies that are conducive to energy conservation and environmental protection, policies limiting the development of high energy-consuming and high polluting industries, and policies promoting the development of industries that conserve energy and protect the environment.[41]

In addition to decreasing per capita GDP energy consumption by 20 percent, the plan seeks to decrease sulfur dioxide and other emissions by 10 percent in order to tackle air and water pollution. In 2008, forty-six new standards were enacted in the law, including twenty-two for energy consumption limits per unit produced, eleven for the end-use energy efficiency of relevant products, five for economical fuel use indicators in the transportation sector, and eight basic standards for energy efficiency metering, computing, and so forth. China also announced fuel economy standards for automobiles, covering all varieties of vehicles, which are more stringent than those in the international community. In total, China has established nearly two hundred energy efficiency standards.

As a result, China is both the world's most rapidly industrializing economy as well as the fastest-growing user of renewable sources of power. In China, policy-driven emissions reductions shake out across four economic sectors: industry, power, buildings, and transportation. The Chinese government has adopted short- and medium-term goals for limiting emissions of heat-trapping gases and a wide-ranging set of policies that contribute to meeting those goals. Those policies are shaped in part by other objectives, including promoting economic growth, cutting local air pollution, and developing strategic industries. It leads the world in solar and wind capacity, adding 43 percent of the world's new renewable power capacity in 2018. Nuclear power and carbon capture technology will also be important in the future, as China strives to balance economic development, resource constraints, and environmental sustainability.[42]

As for transportation, China offers carmakers one of the greatest opportunities of the twenty-first century. In new vehicle acquisition, China passed the United States in 2010 to become the world's largest market for new cars, and, therefore, the American manufacturer Ford is making sure to cater to its unique needs. In a 2010 *New York Times* article titled "In China, Back to Henry's Way," Keith Bradsher and Vikas Bajaj argue that the company's new emphasis on small, inexpensive models (and its scaling back on large, luxury models) is a return to the company's roots. Ford's rise in Asian sales follows the 2010 decision by General

Motors (GM)—which still sells more vehicles in Asia—to focus on the European market and allow its Chinese partner, Shanghai Automotive, to take over its operation in China and India. In 2018, China again led the world in electric vehicle deployment. Roughly 45 percent of the electric cars and 99 percent of the electric buses in the world today are in China. It is clearly a time of significant change in the world's transportation business. And we have not yet discussed possibly the world's most innovative manufacturer of all.

As another indicator of a new era of industrial development, India houses one of the fastest-growing internal combustion engine–powered vehicle manufacturers, Tata Motors. India's market possibilities have been identified by other manufacturers, including Honda, which established centers in India in 2006 with a goal of producing and selling a total of 150,000 units per year by 2010. Instead, Honda has sold nearly double this expectation. The growth of this market is one of the driving forces behind Tata, which already has auto manufacturing and assembly plants in Jamshedpur, Pantnagar, Lucknow, Ahmedabad, and Pune in India, as well as in Argentina, South Africa, and Thailand. A manufacturer of buses, trucks, and trains, Tata has in recent years diversified its auto division in a number of important ways: First, in 2008, as the American manufacturers collapsed and sought to sell off specific brands, Tata purchased Rover and Jaguar from Ford. Second, Tata has set out to make and launch the world's cheapest car: the Tata Nano, priced at 100,000 rupees (US $2,500). Third, rechargeable battery versions of each model began being sold in 2010. Following these initiatives, though, Tata has largely focused inward, experiencing growth only in India. This is a significant market, though. By 2017, India was the fourth-largest market for new autos, surpassing Germany as volume exceeded four million units.

Despite this changing marketplace, Tata has created a new model for manufacturers as, from an early stage, it operates as a truly international corporation. Release of Tata's electric versions will first occur through its UK subsidiary to markets in Finland. Its production and assembly operations now occur in several other countries, including South Korea, Thailand, South Africa, and Argentina, with expansion planned in Turkey, Indonesia, and countries in Eastern Europe. Tata also is developing joint venture assembly operations in Kenya, Bangladesh, Ukraine, Russia, and Senegal and has sales dealerships in twenty-six countries across four continents. Tata has positioned itself to take advantage of this expansive moment, much as Ford did a century ago.

In this new era of personal transportation, manufacturers prioritized flexibility of form and power. Most importantly, though, companies prioritized the flexibility needed to extend their products in a global market, both in terms of the business of manufacturing and distribution, as well as the actual product that was being created. Across the fleet and throughout the world, the automobile provides one of the great indicators of significant change in attitudes toward energy: after decades of aesthetic decadence, the automobile had once again become a utilitarian device for the movement of humans all over the globe.

The emergence of EVs seems to have reached its tipping point in 2021 as each large manufacturer attempts to outdo the others in declaring how quickly they will switch their fleets' production away from the internal combustion engine. Following the American presidential election in 2020, GM rocked the automotive world by proclaiming that it would manufacture only electric-powered vehicles by 2050. Other manufacturers quickly followed with similar proclamations, including Volvo setting its goal for 2035. Toyota set a similar timetable for de-emphasizing gasoline-powered vehicles; however, it was the only manufacturer to emphasize hybrids. The rush into EVs has been spurred by plans in China, Europe, and elsewhere to use government mandates in coming years to stimulate greener forms of transportation in order to assist federal efforts to meet emission targets. The stage is potentially set for a historic cooperative effort between manufacturers and governments to satisfy green consumers.

NOTES

1. CNBC, "Russian and China vie to beat the U.S. in the trillion-dollar race to control the Arctic," https://www.cnbc.com/2018/02/06/russia-and-china-battle-us-in-race-to-control-arctic.html. Accessed January 6, 2022.

2. Defense News, "China's strategic interest in the Arctic goes beyond economics," https://www.defensenews.com/opinion/commentary/2020/05/11/chinas-strategic-interest-in-the-arctic-goes-beyond-economics/. Accessed January 6, 2022.

3. Resilient Cities Network, "Calgary's story," https://resilientcitiesnetwork.org/networks/calgary/.

4. Christopher Jones, *Routes of Power* (Cambridge: Harvard University Press, 2016), pp. 233–235.

5. *Energie.wenden* (Munich: Deutsches Museum, 2016), p. 16.

6. *Energie.wenden*, p. 12.

7. *Energie.wenden*, p. 17.

8. *Energie.wenden*, p. 18.

9. *Energie.wenden*, p. 19

10. *Energie.wenden*, pp. 63–65.

11. William Bogard, *The Bhopal Tragedy* (Boulder: Westview Press, 1989), p. ix.

12. See, for instance, Jack Doyle, *Taken for a Ride: Detroit's Big Three and the Politics of Air Pollution* (New York: Four Walls Eight Windows, 2000).

13. Ross Gelbspan, *Heat Is On* (New York: Basic Books, 1998), pp. 9–13.

14. P. Crutzen, "Geology of Mankind," *Nature* 415, no. 23 (2002). https://doi.org/10.1038/415023a.

15. David Kirsch, *The Electric Vehicle and the Burden of History* (Newark, NJ: Rutgers University Press 1986), p. 205.

16. Kirsch, p. 37.

17. See, for instance, Brad Plumer, "Europe's Love Affair with Diesel Cars Has Been a Disaster," https://www.vox.com/2015/10/15/9541789/volkswagen-europe-diesel-pollu tion; John Vidal, "The Rise of Diesel in Europe: The Impact on Health and Pollution," https://www.theguardian.com/environment/2015/sep/22/the-rise-diesel-in-europe-im pact-on-health-pollution.

18. Brian Black and Gary Weisel, *Global Warming* (New York: Greenwood/ABC-Clio, 2010), pp. 67–70.

19. Black and Weisel, p. 150.

20. Black and Weisel, p. 145.

21. Gelbspan, pp. 9–13.

22. Kenneth Deffeyes, *Hubbert's Peak: The Impending World Oil Shortage* (Princeton, NJ: Princeton University Press, 2001), p. 4.

23. Readers may wish to consult Allan Silverleib's "The Gulf Spill: America's Worst Environmental Disaster?" on CNN.com, for which the author was interviewed: http://www.cnn.com/2010/US/08/05/gulf.worst.disaster/index.html.

24. Helen Campbell, "Discoveries of the Deep," *The BP Magazine* 3 (2007), pp. 46–54.

25. This quote is from news coverage of late May 2010. See, for instance, "Spillcam Shows Gulf Oil Leak Live Online," http://www.upi.com/Top_News/US/2010/05/20/Spillcam-shows-gulf-oil-leak-live-online/UPI-75141274386498/. Accessed on January 6, 2022.

26. See Russell Gold, *Boom* (New York: Simon & Schuster, 2015).

27. Recent coverage includes https://www.nytimes.com/2019/03/26/business/shell -polyethylene-factory-pennsylvania.html. Accessed on January 6, 2022.

28. Adam Rome, *The Genius of Earth Day* (New York: Hill & Wang, 2013), p. 210.

29. When Anwar Sadat urged OPEC members to "unsheathe the oil weapon" in early 1973, the primary rationale for this action was politics.

30. The importance of petroleum to social and economic considerations by the late twentieth century has made its access among the chief concerns of developed nations. See Brian Black, *Crude Reality: Petroleum in World History* (New York: Rowman & Littlefield, 2020).

31. The most useful survey of the evolution of environmental policy in the 1970s is Richard N. L. Andrews, *Managing the Environment, Managing Ourselves: A History of American Environmental Policy* (New Haven, CT: Yale University Press, 1999), especially pp. 225–230; see also Robert Gottlieb, *Forcing the Spring: The Transformation of the American Environmental Movement*, 2nd ed. (New York: Island Press, 1994), as well as Paul Sabin's current work on the approaches to growth in the 1970s. The connection with energy is made in the work of Amory B. Lovins, *Soft Energy Paths* (New York: HarperCollins, 1979).

32. Brian Black, "Oil for Living: Petroleum and American Mass Consumption," *Journal of American History*, special issue on "Oil in American Life," Spring 2012.

33. Brian Black, "The Consumer's Hand Made Visible: Consumer Culture in American Petroleum Consumption of the 1970s," in *Energy in the 1970s*, Robert Lifset and Joseph Pratt, eds. (University of Oklahoma Press, 2014).

34. CNBC, "Led by Tesla, electric vehicle sales are predicted to surge in 2021," https://www.cnbc.com/2020/05/29/led-by-tesla-electric-vehicle-sales-are-predicted-to -surge-in-2021.html. Accessed on January 6, 2022.

35. Audi Green Police advertisement, https://www.youtube.com/watch?v=c9ciAi8e -e0 Accessed on January 6, 2022.

36. Jack Ewing, *Faster, Higher, Farther* (New York: Norton, 2017), p. 146.

37. Ewing, pp. 177–179.

38. Ewing, pp. 196–197.

39. Information drawn from Joel K. Bourne Jr., "Green Dreams: Making Fuel from Crops Could Be Good for the Planet—after a Breakthrough or Two," *National Geographic*, November 2007, at ngm.natiaonalgeographic.com/2007/10/biofuels/biofuels -text.

40. Jad Mouwad, "More Saudi Oil Goes to China than to U.S.," *New York Times*, March 20, 2010.

41. Quoted in Dadi Zhou, "The Process of Sustainable Energy Development in China," Carnegie Endowment for International Peace, August 7, 2009, www.carnegieen dowment.org/publications/index.cfm?fa=view&id=23482.

42. Zhou.

Epilogue

Divining Our Energy Future:
Game Over or Game On?

Our earthly disaster unfolds in real time, attacking our planet's infrastructure often out of human sight or perception. Unlike a nuclear explosion or the impact of an asteroid striking Earth, there is no immediate, existential impact that would lead us to consider obvious reactions such as to pile sandbags against rising waters, to shield our faces from blowing sand and dirt, to duck and cover, or, most simply, to run. Each day we learn new, problematic realities but, by and large, are simply left to continue our everyday patterns of life.

From wildfires to more frequent and violent hurricanes, overt, obvious flash points demonstrate for us our predicament in 2021–2022; however, they are just the most obvious indicators of what goes on below the surface. The true state of things is only available through the use of scientific data or instruments—which, of course, requires us to first accept the findings in order to take on any mitigation efforts. For instance, on May 9, 2013, an instrument near the summit of Mauna Loa in Hawaii recorded a long-awaited climate milestone: the amount of carbon dioxide in the atmosphere there exceeded four hundred parts per million (ppm) for the first time in fifty-five years of measurement—and approximately in three million years of Earth history. For all knowledgeable observers, exceeding 400 ppm was the smoking gun—there was no longer doubt.

You see, the era when CO_2 levels last reached 400 ppm was approximately 2.6 and 5.3 million years ago, which geology refers to as the Pliocene Epoch. Until the twentieth century, Earth's CO_2 level hadn't exceeded 300 ppm—let alone 400 ppm—for at least eight hundred thousand years, which is how far back scientists have been able to measure CO_2 directly in bubbles of ancient air trapped in Antarctic ice cores. Geologists assume that millions of years ago, though, CO_2 must have been much higher than it is now, because Earth was a much warmer place. In the Eocene, approximately fifty million years ago, warm climates and swampy forests such as those in today's southeastern United States reached up into the area off northern Greenland in the Canadian Arctic. It is estimated that CO_2 may have been anywhere from two to ten times higher in the Eocene than it is today.[1] The culprit at that time, of course, was a volcanically active and emerging planet. Today, the influence is not organic. It is artificial. It is us.

Since this fateful moment in 2013, global CO_2 levels—not just those at Mauna Loa—have surpassed 400 ppm.[2] Atmospheric CO_2 levels are directly correlated with rising global temperatures. But it's the warming itself that captures the most international attention among experts. The stark reality of 400 ppm led activists and global policy makers to mitigate living patterns with the aim of keeping the climate from warming more than two (or 1.5 if possible) degrees Celsius, above its preindustrial condition. A special report from the Intergovernmental Panel on Climate Change, released in 2019, estimated that emitting no more than 420 billion to 570 billion tons of carbon dioxide would give the world approximately a 66 percent chance of meeting the 1.5°C temperature goal.

Thanks to the smoking gun of scientific measurement and interpretation, ours is the generation that must reconcile this reality into the requirements of our high-energy existence. Over hundreds of years, human societies have defined the "have" category—societies that would advance and succeed—as increasingly energy intensive. In general terms, beyond mere survival development and advancement was the overall goal for the species, and energy enabled it to happen. In the preceding pages, this book lays out a chronology of energy use that focused first on technology and prime movers of the biological old regime. Advancements in agriculture grew into industrial development that applied stores of energy to innovation and production that altered our species, often raising the standard of living of the "have" populations beyond anything hunter-gatherers could have imagined. This was human progress as generations have identified it. "Haves" sought to advance and maintain their status, while "have-nots" often organized so that they might gain access to energy supplies in order to change their status. Scientists have explained that rising above 400 ppm has largely been a product of our reliance on fossil fuels to provide the energy that many of us have identified as essential and that mitigating Earth's situation requires that to change.

We are left to ponder the rhetorical question: Is this game over or game on?

GAME ON

After accepting this reality, great strides are afoot to provide opportunity for a more sustainable future. For instance, outside of Reykjavik, Iceland, the company Reykjavik Energy runs plants involved in manufacturing aluminum and energy. Even though their plants run on geothermal and hydroelectric power, they give off large amounts of CO_2. In Iceland's effort to become carbon neutral as a nation, these industries are experimenting with capturing the CO_2 released from their smokestacks and injecting it into basalt rock nearby in order for it to turn to stone—basically reversing the geologic process.

Known as carbon capture and storage (CCS), this effort is one of many being tested globally. Typically, carbon capture and storage involves collecting the CO_2 and separating it from other gases, transporting it by pipeline or ship to a suitable site, and then injecting it deep underground. It can be injected into large areas of sedimentary rock or depleted oil and gas fields, among other sites. There it is stored, usually at depths of at least one kilometer, and over time it is turned into a harmless carbonate mineral, such as calcite—one of the main components of marble and limestone.[3] Essentially, carbon can be pulled out of the sky and made into a promising energy source for the future.

The "game on" of fundamentally changing ideas of energy acquisition and generation can be a great equalizer between nations that have and those that have not. And this fact is not lost on developed nations as they step up extreme and expensive efforts to develop remaining fossil fuel supplies. At the heart of these energy initiatives being carried out by China, the United States, Russia, and others lay oil at a stable, moderately high per-barrel price: normally, $50–$60 per barrel prices in recent years. However, the energy market was shaken by a pricing collapse in 2020 with work stoppages caused by the COVID-19 pandemic.

Overproduction of oil in spring 2020 met stymied markets for air and vehicle travel as the vast majority of global humans faced stay-at-home orders. The free fall of prices to $20 per barrel spurred conflicting action from Saudi Arabia and Russia as each nation sought to exploit its own position on the global market. But the upheaval was not over. Per-barrel oil prices plummeted further, breathtakingly reaching negative territory at -$37 per barrel. After a decade in which nations and companies prioritized production, tankers were kept at sea full of crude to hope for an improvement in oil markets. Beyond its symbolic importance, during this moment in oil culture storing crude at sea ballooned to cost oil companies $200,000–$300,000 *per day*. Headlines in 2020 read "Oil Markets Are a Mess. Can World Leaders Straighten Them Out?" and others connected directly to the changing context of "extreme energy production" by proclaiming: "The Coronavirus May Kill Oil Fracking." Nations actively diversifying the sources of their energy supply may find the clearest path forward following the 2020–2021 pandemic.

Indeed, if the pattern of shifting away from petroleum continues and combines with the pressures of peak oil and climate change, those petroleum-filled tankers waiting at sea for a price correction may become a symbol of a different gap: that separating autonomous nations, in control of their own energy futures, and the others who compete for a dwindling supply of finite petroleum. Such a future would seem destined to lead to more and more aggressive wars and conflict regarding the remaining energy supplies.

One can clearly see that in a few years, the new, updated edition of this book will end with a chapter that discusses how, after a century of realizing energy's critical importance to national development and success, a "Great Reversal Redux" may have reformed global power around nations that built an infrastructure organized by an acknowledgment of energy's importance while also more fully accounting for science and sustainability. Chief among such developments will likely be the maturation of climate change from the status of a scientific concept to that of a global problem demanding mitigation. Similar to experiments with carbon sequestration technology, this reality has spurred interest in the construction of a model of energy pricing that will account for the outcomes—particularly pollution—attached to burning all fossil fuels.

Most importantly, emissions from burning fossil fuels such as petroleum are now being factored into commodity costs through "carbon accounting." Informal arrangements for "carbon offsets" are one example; however, much more sweeping examples are now being considered. With the full accounting of fossil fuel energy sources and their impacts on human health, the environment, and climate change, alternative energy sources have become mainstream. This full accounting of the price of fossil fuels can be done in a variety of ways:

- Ideally, the producer of a certain type of energy should be required to pay for its production and all detrimental effects to society and the environment. When this is done, the producer would then pass this cost along to the consumer. The consumer would then be able to reap the financial benefit if they were to choose a low energy–existence life. Even without complete production-side accounting, the government plays an important role in energy accounting by providing incentives to those who use renewable energy and purchase products that consume less energy. These incentives are nearly always financial in nature, so they don't provide for cleaner air or environment.

- Another way for the government to promote a full accounting of energy production is to establish a carbon tax or carbon-trading scheme. The emission of carbon dioxide is the leading cause of global climate change and will have an impact of massive proportions in future generations. By enacting a carbon tax, the government doesn't stop the emission of carbon dioxide and the accompanying climate change, but it does make those emissions more expensive. The producer of energy that emits carbon dioxide then must pass this cost along to the consumer. This is similar to production-side accounting and encourages the use of energy producers who don't emit carbon dioxide.

- A third way for the government to be involved is to pass laws to prevent the emission or release of harmful pollutants. This is sometimes called a "command and control" structure by those opposed to it. With this legal requirement, an energy producer must take necessary steps at whatever costs to prevent the harmful pollution. This cost is then passed on to the consumer. This is production-side accounting. If this were done, it would not be necessary for renewable incentives or carbon taxes to be provided. However, this type of legal requirement to prevent harmful pollution has proven very difficult to enact and enforce.[4]

In practice, the governments of many developed nations already employ a mix of these accounting schemes, and they have had the effect of making alternative energy production more cost-competitive with fossil fuels. As more of these schemes are employed to account for additional harmful pollution from the use of fossil fuels, alternative energy will continue to become more cost-effective and, very likely, the use of petroleum and other fossil fuels will become cost-prohibitive. A new world order could be around the corner in which global political power is enhanced by nations that have emerging technologies such as carbon sequestration to offer a smoother transition in contrast with nations that have not and, instead, continue to rely on the remaining stores of fossil fuels to fill their energy needs. In such a scenario, the paradigm laid out in this book abruptly twists but continues, with energy at the core of human life.

NOTES

1. *National Geographic*, "Climate Milestone: Earth's CO2 Level Passes 400 ppm," https://www.nationalgeographic.org/article/climate-milestone-earths-co2-level-passes-400-ppm/. Accessed on January 6, 2022.

2. http://400.350.org/ and *Scientific American*, "CO2 Levels Just Hit another Record—Here's Why it Matters," https://www.scientificamerican.com/article/co2-levels-just-hit-another-record-heres-why-it-matters/. Accessed on January 6, 2022.

3. BBC, "How Iceland Is Undoing Carbon Emissions for Good," https://www.bbc.com/future/article/20200616-how-iceland-is-undoing-carbon-emissions-for-good. Accessed on January 6, 2022.

4. Brian Black and Richard Flarend, *Alternative Energy* (New York: Greenwood/ABC-Clio, 2010).

Index

EXPLORING WORLD HISTORY

SERIES EDITORS

John McNeill, Georgetown University
Kenneth Pomeranz, University of Chicago
Jerry Bentley, founding editor

Plagues in World History
by John Aberth

Crude Reality: Petroleum in World History, Second Edition
by Brian C. Black

To Have and Have Not: Energy in World History
by Brian C. Black

The World Cup as World History
by William D. Bowman

Getting High: Marijuana in World History, Updated Edition
by John Charles Chasteen

The Age of Trade: The Manila Galleons and the Dawn of the Global Economy
by Arturo Giraldez

The Struggle against Imperialism: Anticolonialism and the Cold War
by Edward H. Judge and John W. Langdon

Smuggling: Contraband and Corruption in World History
by Alan L. Karras

Europeans Abroad, 1450–1750
by David Ringrose

The First World War: A Concise Global History, Third Edition
by William Kelleher Storey

*Insatiable Appetite: The United States and the Ecological Degradation
of the Tropical World, Concise Revised Edition*
by Richard P. Tucker

www.ingramcontent.com/pod-product-compliance
Lightning Source LLC
Chambersburg PA
CBHW030917150426
42812CB00045B/201